BIOLOGY

For Advanced Level

Glenn and Susan Toole

Course Study Guide

D0682387

Stanley Thornes Publishers Ltd

First published in 2000 by:

Stanley Thornes (Publishers) Ltd
Ellenborough House
Wellington Street
Cheltenham
GL50 1YW

00 01 02 03 04 / 10 9 8 7 6 5 4 3 2 1

A catalogue record of this book is available from the British Library.

ISBN 0 7487 3963 7

New Understanding Biology for Advanced Level
ISBN 0 7487 3957 2

New Understanding Biology for Advanced Level plus Course Study Guide
ISBN 0 7487 3964 5

**New Understanding Biology for Advanced Level plus Course Study Guide
(Trade Edition)**
ISBN 0 7487 4467 3

Images of Biology for Advanced Level CD-ROM
Single- and Multi-user Versions available
Free sample available on request
Please contact the publisher at the address given above
or telephone Customer Services on 01242 267273

Acknowledgements
The Publishers are grateful to Patrick McNeill for checking the accuracy of the
information about Advanced level and key skills qualifications.

The Publishers thank the following for permission to reproduce photographs:

Biophoto Associates: pp. 12, 13, 61 (right)
Bruce Coleman (Dr Frieder Sauer): p. 61 (left)
Garden Matters Photolibrary (Colin Milkins): p. 26
Science Photo Library: pp. 12, 13
Science Pictures Ltd: pp. 12, 13, 60

Every effort has been made to contact copyright holders. The Publishers
apologise to anyone whose rights have been overlooked and will be happy to
rectify any errors or omissions at the earliest opportunity.

Typeset by Mathematical Composition Setters Ltd, Salisbury, Wiltshire.
Printed in Great Britain by Ashford Colour Press, Gosport, Hampshire.

Contents

Awarding Bodies

The Associated Examining Board (AEB):

AQA (AEB), Stag Hill House, Guildford, Surrey
GU2 5XJ
Website: www.aeb.org.uk
e-mail: aeb@aeb.org.uk

Edexcel:

Edexcel Foundation,
Stewart House, 32 Russell Square,
London WC1B 5DN
Website: www.edexcel.org.uk
e-mail: enquiries@edexcel.org.uk

Northern Examinations and Assessment Board (NEAB):

AQA (NEAB), Devas Street, Manchester M15 6EX
Website: www.neab.ac.uk/syllabus/maths/
biology.htm
e-mail: biopsy@neab.ac.uk

Northern Ireland Council for the Curriculum Examinations and Assessment (CCEA):

CCEA, Clarendon Dock, 29 Clarendon Road,
Belfast BT1 3BG
Website: www.ccea.org.uk
e-mail: info@ccea.org.uk

For Oxford, Cambridge and Oxford & Cambridge examinations see:

OCR, Publications Department,
Mill Wharf, Mill Street, Birmingham B6 4BU
Website: www.ocr.org.uk

Welsh Joint Education Committee (WJEC):

WJEC, 245 Eastern Avenue, Cardiff CF5 2YX
Website (English language): www.wjec.co.uk
(Welsh language): www.cbac.co.uk
e-mail: bookshop@wjec.co.uk

Introduction

This Course Study Guide has been designed to help you organize your own study. It contains important diagrams from *Understanding Biology for Advanced Level* (4th edition) for you to clip into your notes. These diagrams are arranged so that there is plenty of space for you to add annotations of your own. In addition the guide provides assistance on practical coursework and the skills you will need to develop for answering examination questions. To enable you to understand synoptic assessments, there is a special section explaining their significance. An area where you will need to take responsibility for your own progress is key skills. To help you we have included a summary of what is required for the three main key skills together with some sug-

gestions for how the evidence may be obtained. Remember that there is also a CD-ROM (*Images of Biology for Advanced Level*) available, which provides an interactive opportunity to explore practical aspects of biology in a more detailed and exciting manner. When you first start AS/A-level biology many of the words seem difficult to understand and remember. For this reason we have included a list of the Latin and Greek roots of many scientific words (biological vocabulary); these may be added to your notes for reference throughout the course. By understanding the origin of many biological terms we trust you will be better able to understand and recall them in future.

1 Moving on from GCSE

As a result of the Government's review of post-16 qualifications, changes to AS and A-level examinations have been made for students starting courses in September 2000.

The Advanced Subsidiary (AS) will help you make the necessary adjustments from GCSE because it is assessed at a standard appropriate for candidates who have completed the first year of study of a two-year A-level course, i.e. between GCSE and A-level. It may be taken as a 'stand alone' qualification or, when combined with the second half of the A-level course (known as A2), the AS forms 50% of the assessment of the total A-level. There are three modules/units at AS and a further three at A2. The full A-level therefore comprises six modules/units. The normal pattern envisaged is for candidates to complete AS at the end of their first year of study and A2 at the end of their second year. Alternatively AS and A2 may be taken together at the end of the second year.

If you have studied biology as a separate science for GCSE, you will probably find that many of the topics studied at Advanced level will be familiar to you, but you are now about to explore them in greater depth. Even if you are approaching Advanced level biology from a background of Dual Award Science where the boundaries between biology, chemistry and physics may have been blurred, there will be many areas of study that you will recognize and you should not be at a disadvantage. Some topics, such as biochemistry and cell ultrastructure, will be new to you all but, on the whole, you will cover the same areas as at GCSE but to a depth that enables you to have a much greater understanding of the material.

Few students find the differences in subject content at Advanced level a problem. The same cannot always be said for the study methods and skills required at this level; it may take you a little while to adjust to the different style. AS and A2 Biology courses are all modular, although there is the choice to take the modules/units in stages as you complete them or altogether at the end of your course. However, if you are studying at a school or college you may have little option but to follow the institution's policy in this matter. If you are taking examinations in each unit as you complete it then it is especially important that you come to grips with the new techniques involved as quickly as possible because your first examinations will be upon you quicker than you might imagine. Always ask for help if you are having trouble – you probably won't be the only one in your group to need additional guidance. At Advanced level you will be following fewer subjects than at GCSE, probably up to five subjects. This means that you will be spending at least twice as long studying each subject at school or college. What many students fail to recognize is that the same is true of the time spent outside the classroom. This, too, must increase for each subject, 5 hours per week being a basic minimum for each Advanced level.

One major difference from GCSE is the large body of factual information that you will need to learn. This must be done as you proceed; there is simply too much to be mugged up on at the last minute. Your understanding must be thorough because greater emphasis than at GCSE is placed on the ability to apply general biological knowledge and principles to novel situations, rather than simply being able to recount a previously learned set of facts. The ability to interpret data in a variety of forms is required in order that examiners can test your understanding and not just rote learning. More open-ended questions will appear and you may not be required to give specific answers, but rather to argue a case for and/or against a particular view. At Advanced level, the ability to analyse experiments, data and information critically and to evaluate the accuracy of results and theories is an extension of the basic training given in these skills at GCSE.

In order to develop the necessary attributes, you will need to read widely on all aspects of biology. It is this, above all, which is essential if you wish to obtain a higher grade at Advanced level where a greater range of sources of information must be used. Scientific journals, for example, are invaluable as a means of keeping up-to-date in a subject that is developing as rapidly as biology. Don't forget that important information can also be gleaned from newspapers and television programmes as long as you take a questioning and critical approach to them. Take notes as you read or listen – you never know when they might come in useful.

Practical skills that you developed at GCSE will be an important starting point for the more refined and extensive skills required at Advanced level and you may find that one or more longer term investigations or projects are part of your subject specifications. There is an increasing trend towards the use of mathematical methods, especially statistics, in biology; these demand greater mathematical skill than required at GCSE.

Above all, you must remember that the greater onus on you as an Advanced level student to organize your own work and use your initiative brings with it a far greater need for self-discipline. Advanced level study can be rewarding and stimulating, but it is always demanding and time-consuming, requiring a sustained effort throughout the course and not just in the final few weeks before an examination.

2 | Techniques in Advanced level biology

Guidance for coursework

Coursework is a feature of all Advanced level biology courses. It takes two forms:

- Assessment of a range of practical work carried out during the course.
- Assessment of an extended investigation (project) performed during the course.

Courses usually require either one or the other but, in some cases, a combination of both.

Broadly speaking, practical work has three main components:

- Microscope work
- Drawing specimens
- Experimental investigations

These three aspects are not mutually exclusive but frequently overlap, e.g. an experimental investigation may involve the use of a microscope which, in turn, involves the drawing of specimens.

Skill 1. Microscope work

It is essential to set up the microscope properly in order to obtain a good clear image of what you are looking at. To this end, check that:

- The lenses are clean – wipe with lens tissue as necessary.
- The objective lens is in place – i.e. it is clicked into position.
- The diaphragm is open.

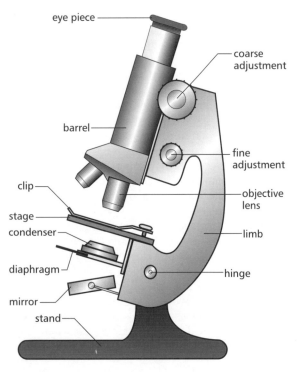

Skill 2. Drawing specimens

Specimens to be drawn for assessment may be:

- Living or dead
- An entire organism or just part of it
- Microscopic or visible with the naked eye
- Real or photographed

Whatever its form, the specimen can be accurately drawn whether or not you are a good artist by following these guidelines:

- Use a good quality, sharp pencil that is not so soft that it smudges, or so hard that you can't rub it out; an HB is usually best.
- Draw on good quality plain paper.
- Choose a suitable scale so that the drawing, labels and annotations will fit comfortably on the page.
- Make large, clear line drawings without the use of ink or coloured pencils.
- Keep the drawing simple by providing only an outline of all the basic structures.
- Draw accurately and faithfully what can be seen. Never draw anything you cannot see, even if it is expected to be present. Never copy from books.
- Draw individual parts of a specimen in strict proportion to each other.
- Provide suitable headings that clearly indicate the nature of the drawing. For microscope drawings, the section (TS/LS, etc.) should be stated.
- State the magnification, scale or actual size of the specimen.
- Label fully all biological features, keeping labels away from the diagram, and never label on the actual drawing.
- Avoid crossing label lines and, if possible, arrange labels vertically one beneath the other.
- Use annotations to indicate the functions of the labelled parts.

Skill 3. Experimental investigations

Experimental investigations test a number of skills:

- Ability to follow instructions
- Construction of suitable hypotheses
- Planning of experiments
- Design and manipulation of experiments

- The stage is properly illuminated – adjust the mirror so that the plane side reflects the maximum light from below the specimen.
- The condenser is correctly adjusted.
- The specimen is in the centre of the stage.
- Low power focusing is carried out before turning to a higher power objective.

- Making accurate observations and recordings
- Making precise measurements
- Presenting results in a suitable form
- Interpreting results accurately
- Making logical deductions from results by applying biological knowledge
- Discussing critically the methods used and results obtained

For much experimental work, you will be given instructions by your teacher, or the examining board, and it is important to adhere strictly to these guidelines. For longer investigations, however, you will usually be expected to plan and carry out your own experiment.

Skill 4. Planning

If planned properly, your investigation should be relatively straightforward. Often the hardest part is deciding what to investigate. To help you make up your mind, consider:

- Extending or modifying an experiment which you have already carried out as part of other work.
- Investigating some biological phenomenon which has intrigued you while studying biology, e.g. some pattern of plant or animal distribution observed during fieldwork.
- Getting ideas from investigations carried out by previous students or eminent scientists.
- Researching textbooks, including *New Understanding Biology for Advanced Level (4th edition)*, which provides useful ideas.
- Choosing something in which you are interested and/or closely involved.
- Keeping it simple – the most basic investigations are often the best.

Having decided what to investigate, you will then need to:

- Predict what will happen in your investigation and form this into a clear and testable hypothesis.
- State and attempt to control all possible variables in the experiment, except those effects you are studying.
- Ensure that your investigation generates quantitative data rather than being merely descriptive.
- Write down all stages of your experiment step-by-step.
- Include references to safety matters and show respect for the living organisms and the environment you study.
- Complete a list of apparatus which you require, including any chemicals. Give **full details**, i.e. state quantities, concentrations, numbers, sizes, etc.
- Make a timetable of what you intend to do over the period of investigation.
- Note on the timetable when you will need specific apparatus or other materials and where this

involves advanced preparation (seedlings do not develop instantaneously from seeds!).
- Make a trial run of the experiment to ensure everything is correct.
- Modify your experiment in the light of your trial run, amending your timetable requirements accordingly.

Skill 5. Writing the method

This should provide a clear, step-by-step account of what you did during your investigation. It should:

- Be written in continuous prose.
- Explain why you proceeded in the way that you did.
- State controls used and the reasons for their inclusion.
- Be presented in such a way that someone else could, by following your account, carry out the same investigation without difficulty and in an identical fashion.

Skill 6. Collecting results

Careful planning and your trial run should ensure that your results can be collected in a straightforward way. Remember to:

- Decide on how many readings you will take and at what intervals.
- Be accurate and precise in taking your readings, e.g. decide to how many decimal places you intend to measure each parameter.
- Record your results in a systematic way on the same data sheet each time.
- Make regular copies of your results as you complete them and store them separately in case the original data is lost.
- Repeat each reading at least once, and preferably twice, to ensure reliable data.
- Make written notes on your results, especially any unusual or unexpected ones, as you go along. This will help in your discussions of the data at a later stage.

Skill 7. Presenting your results

Raw results must be included in your final account of the investigation, but you will also need to present these in a suitably summarized form which allows patterns and relationships to be easily observed. Possible methods of presenting data include:

- Tables of data
- Line graphs
- Histograms
- Bar graphs
- Pie charts

Skill 8. Interpreting results and drawing conclusions

Your results should allow you to accept or reject your original hypothesis, but you will also need to offer explanations and evidence for drawing the conclusions you make. Your interpretation and discussion should therefore include:

- A statistical test of significance applied to your results, e.g. χ^2 test, t-test.
- Explaining trends and patterns demonstrated by results.
- Applying theoretical knowledge in interpreting and explaining results.
- Identifying and accounting for results which do not fit the trend.
- Identifying possible sources of error in the results.
- Suggesting how these errors might be minimized or overcome. (It is important to restrict yourself to errors which were beyond your immediate control. It is not a good idea to highlight your own inadequacies!)
- Discussion of the biological importance of your findings.
- Reference to relevant published work and/or research you carried out, e.g. letters to commercial and professional organizations.

Skill 9. Writing your final account

It is wise to adopt the accepted style used in the publication of scientific papers, namely:

- A title.
- An abstract, i.e. a short account of the hypothesis being tested, methods used and overall conclusions.
- An introduction explaining the hypothesis and referring to similar work carried out on the topic.
- The method, written in continuous prose.
- The results, in summarized format.
- A discussion of the results and their significance.
- A bibliography quoting all references used in the study.
- An appendix which contains your plan, raw results, statistical tests and relevant background material. This often lengthy information should be added at the back and be well organized and clearly labelled in sections so that it is easily accessed through appropriate cross-referencing in the main text of the investigation.

3 | Tackling examination questions

The style of examination questions set in AS/A-level biology examinations has become increasingly standardized over recent years. The majority are now structured questions that are divided into separate small parts and are answered in the space provided on the question sheet. These questions vary considerably in length and in the degree of analysis required. For example some are tables in which ticks and crosses must be placed appropriately while others may require graphs to be drawn or interpreted, statistical problems to be solved or comprehension passages to be attemped. It is also increasingly common to come across longer prose or essay type questions, especially in A2 units. Multiple choice questions are no longer set.

Structured (short-answer) questions

There is great variation in this type of question. In its simplest form, only a single word answer is required, for example 'Name the enzyme that hydrolyses starch to maltose'. Some demand a short sentence, for example 'What is the function of follicle stimulating hormone?', while others require a longer response, for example 'Describe how you would carry out Gram's staining technique on a sample of bacteria'. These types of question may require you to demonstrate any of the following skills:

- Show knowledge and understanding of biological terms, concepts, principles and relationships.
- Construct hypotheses.
- Design experiments.
- Interpret the results of experiments.
- Draw conclusions and make inferences.
- Assess and evaluate numerical and non-numerical information.
- Explain observations and solve problems.
- Present data in its many varied forms.
- Comprehend, interpret and translate data.
- Criticize material and exercise biological judgement.
- Construct or label diagrams of biological importance.
- Interpret or comment on photographs.
- Collect, collate and summarize biological information in an appropriate form.
- Appreciate the social, environmental, economic and technological applications of biology.

In structured questions, it is content rather than style that is being tested and so answers should be clear and concise and to the point; note-style responses are acceptable. The marks allocated and the space provided for the answer usually give an indication of the amount of detail required.

Essay (free-response) questions

Most Awarding Bodies set at least one question that requires you to produce a longer answer, often in the form of a short essay. Being open-ended, such questions do not limit you to one particular response, but rather give you the scope to demonstrate the depth and breadth of your biological knowledge. It is often

for this reason that students dislike them – they are uncertain of what exactly the question requires and/or lack the range and detail of knowledge needed to answer them effectively. Equally, there are some who fare badly on essay questions for the opposite reason – they digress widely from the question, using it as an opportunity to relate the last biological article they read or indulge their opinions with little regard to their relevance. The skill lies in achieving a compromise between these two extremes.

Essay questions may be structured to some degree. In a highly structured essay, the question comprises a number of different parts that often guide you step-by-step through the question, making clear at each stage exactly what is required. To take a typical example:

a) Explain what is meant by the terms:
   (i) enzyme
   (ii) coenzyme

(4 *marks*)

b) With reference to the lock and key theory, and active sites, explain how an enzyme works.

(8 *marks*)

c) What are the factors that influence the rate of enzyme activity?
   Briefly point out why the factors you mention alter the rate.

(8 *marks*)

The figures in brackets represent the marks allocated to each section – a common feature of the structured essay. These marks provide clear guidance on how long should be spent on each part. In this example, it is relatively clear what the examiner is looking for at each stage. The answer to this question might not be very different from one entitled 'Write an essay on enzymes', but we suspect that you would find the latter more problematic because you cannot be sure what the examiner wants and therefore what to include and what to leave out. The structured essay overcomes these problems to some degree, but you should never neglect careful planning. Under each subsection of the question, it is necessary to jot down the essential points to be made before starting your answer. Only in this way can your response be logically argued with well-marshalled ideas and relevant supporting evidence. You should answer each part separately – never attempt to merge them into a general essay.

Take care not to use information more than once as you cannot be given credit twice for the same explanation. In the question cited it would be feasible to include the lock and key theory in all three parts. This would be a waste of examination time. It is almost certainly to avoid this problem that the examiners have guided you to include this information in part (b).

An alternative type of essay is the unstructured one. This may be as open ended as 'Write an essay on enzymes/carbohydrates/animal reproduction/ photosynthesis/hormones', etc. In practice, they are often a little more explicit:

- Describe the properties of water and show why it is vital to living organisms.
- Describe how animals depend for their survival on the activities of plants.
- Argue a case for and against the use of pesticides and fertilizers.

In this style of essay, planning is more crucial. The number of points that could be made is vast and the skill is limiting them to the most important, i.e. those most likely to be on the examiners' mark scheme. In the 30–40 minutes typically allowed for an essay question, only perhaps 20 or 30 individual facts can be included in sufficient detail to warrant marks. You must ensure that these points cover the full range of the topic being examined. Take, for example, the following essay: 'Write an essay on movement in organisms'.

There is a danger here that all points made will be biased towards the movement of whole organisms (locomotion) and animals. Good planning (and adequate knowledge) can avoid this. Movement in biology includes the movement of material within an organism and within and between cells. All living things use movement. Your essay should, therefore, encompass all types of movement and a wide range of organisms – plant, animal and Protoctista. Avoid any bias towards animals, especially mammals.

The main areas on your plan should be:

Types of movement	Example
Subatomic movement	Electron movement in photosynthesis
Atomic/molecular	Diffusion, osmosis, active transport into and out of cells, transpiration, translocation
Cellular	Amoeboid, ciliate, flagellate locomotion; movement of macrophages
Tissue/organ	Heart pumping, circulation of blood
Organism	Locomotion, swimming, flying, burrowing, etc.
Population	Migrations

Include each of the six areas listed, but try to link them in a coherent manner. Some Awarding Bodies give marks for the manner in which the essay is written, its fluency, clarity and logical arguments.

The wording of essay questions is all important. There are subtle differences between terms such as 'compare', 'describe', 'discuss', and 'distinguish'.

Unless you appreciate this, you are likely to needlessly throw away marks. The following list is a guide to the appropriate meaning of a number of commonly occurring question terms.

Describe	Give an account of the main points, with reference to (visual) observations if possible.
Recall	Identify and reinforce knowledge gained at key stage 4 or from other sources such as other units in the specification.
Explain/Account for	Show how and why; give reasons for – with reference to theory, if possible.
Compare	Point out differences and similarities.
Distinguish/Contrast	Make distinctions between, recognize comparable differences.
Discuss	Debate, giving the various viewpoints and arguments, in a balanced, reasoned and objective manner.
Criticize	Point out faults and shortcomings.
Survey	Give a comprehensive and extensive review.
Comment on	Make remarks and observations on.
Understand	Explain underlying principles and apply knowledge to new situations.
Appreciate	Demonstrate awareness of the significance but without detailed knowledge.
Illustrated	Use figures, drawings, diagrams.
Annotate	Add notes of explanation.
Calculate	A numerical answer is required and working should be shown.
Briefly/Concisely	Give a short statement of only the main points.
Outline	Give the essential points, briefly.
List	Catalogue, often as a sequence of words one beneath the other.
Significance	Show the importance of.
State	Set down concisely with little or no supporting evidence.
Define	Give only a formal statement.
Suggest	Put forward ideas, thoughts, hypotheses.
Devise	Construct, compose, make up.
Estimate	Give a reasonable approximation.

Modular tests

Your A-level course will involve you taking a series of modular tests, even though these may be all taken together at the end of the course.

Preparation for modular tests and the examination techniques you need to apply to them are largely the same as for other written examinations. However, the following points should be remembered when taking modular tests.

- The modular tests, even when taken after only one term of study, are marked to the same standard as terminal tests. For AS papers this means a standard between GCSE and A-level. For A2 papers you are expected to reach full A-level standard.
- The modular tests will cover all the material of that module and therefore examine a narrower range of information than the synoptic test taken at the end of the course. Do not be lulled into thinking that preparation can be less thorough and/or less urgent, especially as all questions are likely to be compulsory.
- The time interval between the practice (mock) test and the actual examination is likely to be short, leaving little time to compensate for deficiencies in knowledge and/or technique.
- It is a requirement of all A-level specifications that 20% of the assessments involve synoptic components (see Section 6).
- Modular tests are typically $1-1\frac{1}{2}$ hours in duration. Their relative shortness makes the management of your time especially important as there is little opportunity to compensate for lost time. Answers need to be concise and to the point.
- There is little respite from examinations, no sooner is one complete than the next appears on the horizon. Take care not to become fatigued, but to adequately build yourself up mentally for the next module.

4 | Model answers to sample questions

Provided you have prepared an adequate revision timetable, begun the process early and followed it through faithfully, there should be little left to do on the evening prior to your examination. If you have not revised adequately it is a mistake to imagine that a few hours of cramming can compensate for your earlier omissions.

The best advice we can give for the evening prior to an examination is to try to relax and have a good night's rest. You may find it difficult to follow this advice as it demands great confidence that your revision has been complete. If you simply cannot bring yourself to do no work, we would suggest a quick skim through your notes or textbook, perhaps reading headings and sub-headings, to generally absorb ideas and principles. Equally, there may be a few equations, definitions or mnemonics that you might wish to reinforce in your memory. Avoid, however, detailed revision of topics as it is almost certainly too late for this to be of benefit and there is the very real risk of inducing panic as you struggle to come to grips with a difficult concept. This will only create confusion and undermine your confidence, making matters worse than if no work had been done at all.

Our advice is therefore to keep any revision low key, to prepare the necessary examination materials (pens, pencils, calculator, etc.) for the following day, relax, set your alarm, and get a good night's sleep.

 ## Final preparations. Practice questions

To help you understand a little of the examiner's mind, we have provided two sample questions below, followed by a guided response to help you see how an answer should be constructed. Tutorial tips are also given to improve your skills in answering questions.

Sample question 1

a) The energy flow in a simple food web of five organisms, A, B, C, D and E is shown below:

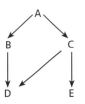

If organism C were suddenly removed from the food web describe, with reasons, the extent to which the populations of the other four organisms might be affected over time.

(12 marks)

b) The distribution of four species of organism at different depths in a pond was investigated and the data presented graphically as shown below:

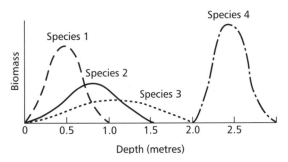

(i) Giving your reasons, state which species is most likely to be the main primary producer.

(ii) Which species is most likely to be a secondary consumer? Give reasons for your choice.

(8 marks)
(Total 20 marks)

Guided response to sample question 1

a) Given that the question states 'the extent to which the populations ... might be affected over time', be sure to consider the **degree** of change **over a period**. As it is impossible to be certain precisely what might happen as we do not know the exact species involved, it is fair to assume that credit will be given for any reasonable response that can be supported by accepted ecological principles. In other words, there may be no single correct answer.

> *Hint*
> To avoid confusion for yourself and the reader, deal with each organism under a separate heading where possible. In this case, however, it is better to deal with A and B together as they closely affect one another.

Organisms A and B

The point here is that organism A is consumed by organism C and therefore:

✓ The population of A will initially increase due to a reduction in the number of its consumer – organism C.

> *Hint*
> Throughout your answer be sure to consider the **indirect**, as well as the **direct**, effects of the removal of organism C from the web.

However, organism B also feeds on organism A, and therefore:

✓ The increase in population A will mean a larger food supply is available for organism B, whose population will probably increase as a result.
✓ With a larger population of B consuming organism A, the latter's population will, in turn, decrease after a while
✓ until a new equilibrium is established between the populations of A and B, at which point both populations will stabilize
✓ (BONUS) probably both at a level higher than the original because the total energy of the system is now shared between fewer species, some of which can, therefore, sustain a larger population.

This response assumes that these five organisms constitute the whole ecosystem, a highly unlikely situation but, in the absence of other information, this is the only way to view it.

> **Hint**
> Always look to extend an argument/explanation to the full – making predictions as necessary. This can often earn valuable bonus marks.

Organism D
Organism D feeds on organism C (which has disappeared) and organism B (whose population has increased). Therefore:

> **Hint**
> Further marks are not available for points already made, so do not repeat here the reasons why population B has increased.

✓ the population of D will initially decrease
✓ because in the absence of organism C, its food supply is diminished
✓ but as the population of B later increases, so more food will be available to replace that lost as a result of organism C disappearing
✓ and so the population of D will later recover.
✓ The extent of this recovery depends on the degree to which the increase in the population of B can compensate for the loss of population C.

> **Hint**
> Assume that the organisms shown represent the **whole** of the food web and do not attempt to consider the effects of other, hypothetical ones, which might also exist.

Organism E
Assuming there is no other food source than organism C on which organism E can live, then:
✓ the population of E would decrease
✓ ultimately becoming completely extinct.
✓ The timescale for this extinction depends upon the

internal and external food stores available to organism E.

b) (i) The two important points to bear in mind when answering this question are the **depth** in the pond at which each species occurs and the **biomass** of each species. Remember that photosynthesis needs light and this is most available near to the surface of the pond, and rapidly diminishes the deeper one gets. Remember, also, that primary producers as a whole have greater biomass than all primary consumers. Considering these points, your response to part (i) might be:
✓ Species 1 is the primary producer
✓ because primary producers are photosynthetic, requiring light to make their food.
✓ As light is more readily available at the surface of a pond, primary producers are more likely to be present in the upper levels of the pond and species 1 is found within the top metre, with a maximum biomass at 0.5 m.
✓ Furthermore, species 1 has a large biomass (although not the largest), which is typical of primary producers.

> **Hint**
> It is the **area** under each curve that is a measure of total biomass, not the maximum height of each curve.

> **Hint**
> Use specific figures from the data provided to illustrate your points and so support your answer.

(ii) In answer to this question, you need also to consider the distribution of each species. Consumers, feeding as they do off another organism, must, at some point, be found at the same depth as the species they feed upon – how else could they consume them if they were never in the same place? The curve for the secondary consumer must, therefore, overlap with that of the primary consumer. This logic would appear to discount species 4, which does not occur at the same depth as any other, besides which its biomass is the largest of those shown, making it highly unlikely to be a secondary consumer. Your answer might appear:
✓ Species 3 is likely to be the secondary consumer
✓ as it has the smallest biomass of the four species shown and,
✓ since energy is lost at each stage in the food chain, so the biomass usually (but not always) reduces as one moves up the trophic levels.
✓ Species 3 is found at the same depth as species 2 (their curves overlap), which is likely to be the primary consumer, as its

biomass is intermediate between species 1 (producer) and species 3 (secondary consumer).

What, then, you might ask, is species 4? Given its location at the lowest depth, and its large biomass, it is almost certainly a decomposer or detritivore. It feeds on dead and decaying remains that fall to the bottom – (hence its location) and it has a large biomass because it can feed on **all** the other species in the pond.

✓ A well-argued elimination of species 4 might well earn a bonus mark.

Sample question 2

In an experiment, the same quantity of the hormone gibberellin was applied to either the first leaf or the stem of dwarf bean plants and the leaf area of the plants was then measured over the following 3 weeks.

The experiment was then repeated under exactly the same conditions except that the tip of each plant was removed at the same time that the gibberellin was added. The control experiment in both cases was to use a group of plants to which no gibberellin was added. The results are shown in the graphs below:

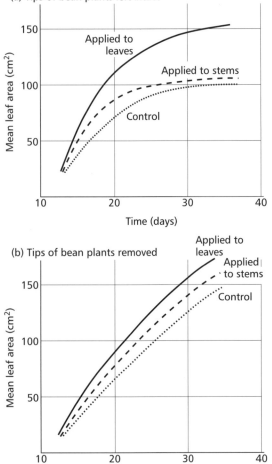

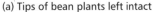

a) (i) Using evidence from the graphs, state what effects the removal of the growing tip has on the growth of the leaves in the absence of gibberellin.

(3 marks)

(ii) From your knowledge of plant hormones in general, give a possible reason for these effects.

(1 mark)

b) What differences occur when gibberellin is applied to the leaves rather than the stem of intact bean plants?

(4 marks)

c) (i) In what ways did the removal of the growing tips influence leaf size when gibberellin was applied?

(5 marks)

(ii) Put forward a hypothesis that could explain these influences.

(3 marks)

d) At a cellular level, how might an increase in leaf area occur?

(2 marks)
(Total 18 marks)

Guided response to sample question 2

a) (i) In the absence of the growing tip, the control experiment (dotted) line rises more steeply, reaches a higher point after 35 days and does not reach a plateau. These three observations lead to the following three conclusions. The removal of the growing tip causes:

✓ slower/less leaf growth/expansion over the initial period of the experiment (approximately 60 cm^2 leaf area after 20 days, compared with 70 cm^2 when the tip remains intact);

✓ more overall leaf growth/expansion over the period of the experiment (approximately 130 cm^2 leaf area after 30 days compared with 100 cm^2 when the growing tip is intact);

✓ leaf growth/expansion continues rather than ceasing after 35 days as it does when the tip remains intact.

> **Hint**
> The control experiment is the one that is carried out 'in the absence of gibberellin'; therefore compare the two dotted lines.

> **Hint**
> Use specific data from the graphs to support your answers.

(ii) It could be argued that the removal of the tip leaves more nutrients available for the rest of the plant, including the leaves, which therefore grow more rapidly, but the phrase 'from your knowledge of plant hormones in general' suggests another reason is required, namely:

✓ Auxins are produced in the growing tips and these inhibit/suppress leaf growth/expansion.

b) Again, you need to observe three differences between the leaf (solid) line and the stem (broken) line; namely that for the leaf, the final point of levelling off (plateau) is higher, the gradient is greater throughout, and the line is still rising (albeit very slightly) at the end of the experiment.

It is also necessary to note that the control (dotted) line and the stem (broken) line meet after approximately 32 days. From these observations, you should conclude that when gibberellin is applied to the leaves rather than the stem, then:

✓ The overall area of the leaves is greater (150 cm^2 compared with 100 cm^2 when applied to the stem);

✓ The **rate** of leaf growth/expansion is greater;

✓ Leaf growth/expansion is still continuing after 35 days, rather than reaching a maximum after 32 days;

✓ The final leaf area is greater than when no gibberellin is applied (control) whereas it is the same if gibberellin is applied to the stems.

c) (i) Here you need to compare the leaf (solid) lines and the stem (broken) lines on the two graphs. This comparison should lead you to state that when the growing point is removed:

✓ The growth/expansion of the leaves is initially slower (up to 20–25 days).

✓ The growth/expansion continued for longer.

✓ The growth/expansion did not level off (plateau).

✓ The final leaf area was greater.

✓ The removal of the growing tip had a much more pronounced effect when gibberellin was applied to the stems than when it was applied to the leaves.

(ii) Any reasonable hypothesis is likely to bring credit, but the most probable one is:

✓ Auxins are made in the growing tips;

✓ and these inhibit/suppress the effect of gibberellin overall;

✓ and may increase the transport of gibberellins into the leaves from the stems.

d) There are two ways in which a group of cells might increase their area/volume:

✓ by increasing their number through mitotic division;

✓ by increasing the size of each cell through expansion/elongation as a result of the uptake of water/nutrients (or a combination of both!).

5 A-level grade criteria

The criteria tabulated below help to explain what Awarding Bodies (examination boards) expect you to be able to do in order to achieve a particular grade. These criteria help to maintain fairness from year to year and between different Awarding Bodies as well as providing a framework that can be understood by students, universities and employers.

Grade E	Grade C	Grade A
Knowledge and understanding Recall and use knowledge from parts of the course Show some understanding of principles beyond that expected at GCSE level Select information to answer structured questions Use a limited range of biological terminology correctly	**Knowledge and understanding** Recall and make sound use of knowledge from many parts of the course Show understanding of some basic principles Select appropriate information and present ideas clearly Use correct biological terminology	**Knowledge and understanding** Recall and use knowledge from most of the course, with few significant omissions Show good understanding of biological principles and concepts Select appropriate technical terms and use them to express ideas
Application Carry out straightforward calculations with some guidance Solve problems concerned with familiar material Where problems concern unfamiliar material, answers are relevant even if difficulties arise in applying facts and principles	**Application** Carry out a range of calculations with minimal guidance Apply basic principles to new contexts Link together ideas from different areas of the course Give correct, relevant and logical answers	**Application** Carry out unaided a range of calculations methodically and explain solutions logically Have a good understanding of biological principles and apply them to new contexts Show insight into problem solving and suggest a number of possible solutions using knowledge from a range of sources Give answers that are correct, coherent, relevant and detailed
Experimental activities Plan experiments with guidance Obtain appropriate results for familiar practical procedures Present tables or graphs of results and interpret broad trends Draw basic conclusions from results and, with assistance, relate these to biological knowledge and understanding	**Experimental activities** Produce an experimental plan, although it may need modification Use a range of techniques to obtain accurate results Explain results and, with help, evaluate them Comment on data and use it to support a particular hypothesis	**Experimental activities** Produce a clear and accurate plan without assistance Use a range of techniques with skill, care and precision Record and interpret data Evaluate techniques and data critically Present clear and concise arguments by weighing up the evidence

6 | Synoptic assessment

In both AS and A2, you will be assessed on:

1. **Knowledge with understanding**

   You should be able to:

   - recognize, recall and show understanding of specific biological facts, terminology, principles, concepts and practical techniques;
   - draw on existing knowledge to show understanding of the ethical, social, economic. environmental and technological implications and applications of biology:
   - select, organize and present relevant information clearly and logically, using appropriate vocabulary where appropriate.

2. **Application of knowledge and understanding, analysis, synthesis and evaluation**

   You should be able to:

   - describe, explain and interpret phenomena and effects in terms of biological principles and concepts, presenting arguments and ideas clearly and logically, using specialist vocabulary where appropriate;
   - interpret and translate from one form into another, data presented as continuous prose, or in tables, diagrams, drawings and graphs;
   - apply biological principles and concepts in solving problems in unfamiliar situations including those which relate to the ethical, social, economic and technological implications and applications of biology;
   - assess the validity of biological information, experiments, inferences and statements.

3. **Experiment and investigation**

   You should be able to:

   - devise and plan experimental and investigative activities, selecting appropriate techniques;
   - demonstrate safe and skilful practical techniques;
   - make observations and measurements with appropriate precision and record these methodically;
   - interpret, explain, evaluate and communicate the results of experimental and investigative activities clearly and logically using biological knowledge and understanding and using appropriate specialist vocabulary.

4. **Synthesis of knowledge, understanding and skills**

   You should be able to:

   - bring together principles and concepts from different areas of biology and apply them in a particular context, expressing ideas clearly and logically and using appropriate specialist vocabulary;
   - use biological skills in contexts which bring together different areas of the subject.

This fourth assessment objective is known as synoptic assessment, and the units/modules which contain it must be taken at the end of the course. Some students find the idea of the synoptic assessment daunting because it does require rather different preparation from that employed for the other modular papers. Usually a module has a very clearly defined content which you will have studied, understood and learnt so that you can answer the questions set on the relevant examination paper. There is no such clearly defined content for the synoptic papers. Instead you will be tested on your ability to recall general principles and apply them to different areas of biology, expressing yourself clearly and using appropriate specialist vocabulary. It is therefore essential that you develop a general feeling for and understanding of your subject during the two-year A-level course. Simple recall of isolated facts will not be sufficient to get over this final hurdle. It is at the synoptic stage that students who have read widely and really developed an all-round interest will come into their own. They will readily see the links required to answer over-arching questions such as

> 'Write an essay on the maintenance of water and solute balance in plants and animals'
> or
> 'Explain the interactions of plants and animals in maintaining a constant composition of the atmosphere'.

Do not think of the synoptic assessment as simply something to be overcome at the end of your course. It is essential that you start to prepare for it early on. Ask questions, discuss issues, read journals and books and keep up to date. Remember that although biology is a science you will still be assessed on the quality of your written communication. When you have to give an extended written answer, such as an essay, you will be expected to:

- select and use an appropriate form and style of writing
- organize relevant information clearly, using specialist vocabulary when appropriate
- ensure the text is legible and that spelling, punctuation and grammar are accurate.

7 | Key skills through Advanced level biology

What are key skills?

Key skills are those fundamental skills which are valuable regardless of the subject you are studying or the career you wish to follow. As such they will help you to improve your own learning and performance whether as part of your education and training, your work or everyday life.

Key skills include:

- communication skills
- information technology skills
- application of number skills
- improving your own learning performance
- working with others
- problem solving

Why are key skills important?

Key skills are important:

- **In your learning** because they help you to achieve better results through assisting you to look at how you learn, and to decide what methods are effective for you.
- **In obtaining employment or continuing your education** because employers and university admissions tutors know how valuable key skills are in helping their staff/students perform well. They therefore look for key skills when recruiting suitable people. So much so that key skills qualifications will gain UCAS points that count towards the entry requirements of university courses.
- **In your career** because key skills are relevant in all careers and at all levels. They are transferable skills which go with you when you move to another post in an organisation or change career altogether. Key skills therefore give you greater flexibility – something

managers look for when deciding who should gain promotion.
- **In your everyday life** because key skills give you the communication, numeracy and organizational abilities to get on well with other people, manage your affairs and handle the vast amount of information now available to us all, not least through the Internet.

How can I obtain a key skills qualification?

You can gain a qualification as part of your post-16 study. Key skills are assessed at different levels. The most appropriate to an AS or A-Level subject using a combination of independent assessment and a portfolio of evidence collected from your day-to-day studies, employment or other activities. Each key skill will lead to its own separate certificate but you will need to succeed in all of the first three (communication, information technology and application of number) to obtain the key skills qualification.

How can I demonstrate key skills at level 3 during my AS/A-Level biology course?

All the key skills can be demonstrated through AS/A-Level biology although you should not expect to cover all of them exclusively through this subject. To help you select which aspects you want to demonstrate through your biology we have listed the skills needed to obtain the level 3 qualification along with suggestions of how they can be achieved through the medium of biology.

Communication Level 3	
You must:	**Example**
C3.1a Contribute to a group discussion about a complex subject. Evidence must show you can: • make clear and relevant contributions in a way that suits your purpose and situation; • listen and respond sensitively to others, and develop points and ideas; and • create opportunities for others to contribute when appropriate.	There will be many opportunities for discussion during your course. For instance about health matters (e.g. diet and coronary heart disease or ageing) about the effects of humans on the environment (e.g. deforestation or air pollution) about DNA technology (e.g. genetically modified organisms) or about ethical issues (e.g. in-vitro fertilization or surrogacy).

	You will need to:
	<ul><li>gather your thoughts before speaking;</li><li>speak clearly and keep to the point;</li><li>use relevant and specialist vocabulary;</li><li>encourage others to have their say and listen sensitively to their opinions.</li></ul>
C3.1b Make a presentation about a complex subject using at least one image to illustrate the complex points. Evidence must show you can:<ul><li>speak clearly and adapt your style of presentation to suit your purpose, subject, audience and situation;</li><li>structure what you say so that the sequence of information and ideas may be easily followed; and</li><li>use a range of techniques to engage the audience, including effective use of images.</li></ul>	You will almost certainly have an opportunity to make a presentation to the rest of your class or another group of people. This may be on a specific topic you have researched or a more open-ended issue such as genetically-modified foods, food additives or abortion. You will need to:<ul><li>Speak clearly and vary the volume and tone of voice. It is a good idea to practise your presentation out loud to get the timing right and if you can find someone to listen to you, he or she may make helpful observations about whether you can be heard, and whether your context makes sense.</li><li>Structure your arguments logically (make notes but don't just read out a prepared script).</li><li>Use one or more images. These could be slides, handouts or overhead projector transparencies. If possible use a computer presentation package such as Powerpoint (thus hitting IT targets too).</li></ul>These will help to keep you on track. You may base your image on a photocopied or scanned image that you have amended in some way, but whatever you do make sure it is relevant and not just decorative.
C3.2 Read and synthesize information from two extended documents about a complex subject. One of these documents should include at least one image. Evidence must show you can:<ul><li>select and read material that contains the information you need;</li><li>identify accurately, and compare, the lines of reasoning and main points from texts and images; and</li><li>synthesize the key information in a form that is relevant to your purpose.</li></ul>	Synthesize means put together'. You could achieve this skill through writing an essay, through preparing information for skill C3. 1 (above) or in planning an individual practical investigation. You will need to:<ul><li>Select relevant information from more than one source. (Use different textbooks, science publications such as *Biological Sciences Review* or *New Scientist*, the Internet, e-mail, etc.).</li><li>Put together relevant information in a suitable form for your purpose.</li><li>Use and interpret visual information – for example a graph or diagram.</li></ul>
C3.3 Write two different types of documents about complex subjects. One piece of writing should be an extended document and include at least one image. Evidence must show you can:<ul><li>select and use a form and style of writing that is appropriate to your purpose and complex subject matter;</li><li>organize relevant information clearly and coherently, using specialist vocabulary when appropriate; and</li></ul>	You could demonstrate this skill when planning and writing your individual study topic, writing an essay or reporting on some practical exercise such as succession in an ecosystem. You will need to:<ul><li>Decide the best style for each purpose – for example practical instructions may best be presented as a series of numbered or bulleted points rather than paragraphs or prose. Diagrams or other images will almost always be needed in biology – especially in practical instructions.</li></ul>

• ensure your text is legible and your spelling, grammar and punctuation are accurate, so your meaning is clear.	• Ensure your work is legible (use a word processor if you can), accurately and correctly punctuated and spelled (use a spell-checker if possible).

Information technology Level 3

You must:	Example
IT3.1 Plan and use different sources to search for, information required for two different purposes. Evidence which must be gained in the course of a substantial activity must show you can: • choose appropriate sources and techniques for finding information and carry out effective searches; and • make selections based on judgements of relevance and quality.	You could achieve this skill when preparing an abstract for an individual study, in researching for a discussion, preparing a presentation or planning a practical exercise. You will need to: • Plan how to obtain and use the information required to meet the purpose of your activity. • Use the Internet using an appropriate search engine. Sites run by a university or the Institute of Biology are likely to be more reliable than the websites of unknown individuals. • Use your judgement in deciding whether the information is complete and unbiased. While Friends of the Earth may stress the conservation aspects of oil exploration, petroleum producing companies are unlikely to do so.
IT3.2 Explore, develop and exchange information to meet two different purposes. Evidence must show you can: • enter and bring together information in a consistent form, using automated routines where appropriate; • create and use appropriate structures and procedures to explore and develop information; and • use effective methods of exchanging information to support your purpose.	This skill could be demonstrated through the use of data-loggers, spreadsheets and secondary data (e.g. on the products of CO by yeast). Data from experiments, such as those with enzymes, could be shared between students (using e-mail if possible) and converted into a graph. Early drafts could be e-mailed to your tutor for feedback. You will need to: • use automated routines such as icons and macros; • sort and group the information to allow conclusions to be drawn; • exchange information between yourself and others.
IT3.3 Present information from different sources for two different purposes and audiences. Your work must include at least one example of text, one example of images and one example of numbers. Evidence must show you can: • develop the structure and content of your presentation using the views of others, where appropriate, to guide refinements; • present information effectively, using a format and style that suits your purpose and audience; and • ensure your work is accurate and makes sense.	You could produce a report using a word-processor on a practical investigation or individual study. You will need to: • include diagrams from a graphics package; • process results using a spreadsheet, charts or graphs; • ensure the final document is proof-read and in a form that is easily understood by the person reading it.

Application of number Level 3	Example
You must	
N3.1 Plan and interpret information from two different types of sources, including a large data set. Evidence which must be gained in the course of a substantial and complex activity must show you can: • plan how to obtain and use the information required to meet the purpose of your activity; • obtain the relevant information; and • choose appropriate methods for obtaining the results you need to justify your choice.	This could relate to the planning of practical work and treatment and calculation of results – using paper and pencil, calculator or computer (for example a spreadsheet). If you take results from the whole of your teaching group you could assemble a large data set. Also you could compare your results, obtained from a small number of cases, with a database of other people's results. Suitable activities could include the use of an eyepiece graticule, making stomatal measurements and carrying out enzyme experiments. You will need to: • pool data, e.g. from others in your group; • select the best method for obtaining your results; • say why you chose this method.
N3.2 Carry out multi-stage calculations to do with: A amounts and sizes; B scales and proportion; C handling statistics; D rearranging and using formulae. You should work with a large data set on at least one occasion. Evidence must show you can: • carry out calculations to appropriate levels of accuracy, clearly showing your methods; and • check methods and results to help ensure errors are found and corrected.	There are many opportunities to demonstrate these skills, especially when dealing with practical results. Amounts and sizes are significant when making microscope drawings especially using an eyepiece graticule. Scales and proportion are used in experiments such as the effects of enzyme concentration on the rate of reaction, or where the different concentrations are obtained by serial dilution. This technique is also used in the plating of micro-organisms to find their concentration. Handling statistics is commonplace in genetics and ecology field work where chi-squared tests, t-tests and other statistical techniques are used. You will need to: • take care to be accurate especially in transcribing results and making calculations; • check over your work to eliminate errors.
N3.3 Interpret results of your calculations, present your findings and justify your methods. You must use at least one graph, one chart and one diagram. Evidence must show you can: • select appropriate methods of presentation and justify your choice; • present your findings effectively; and explain how the results of your calculations relate to the purpose of your activity.	Many of your practical write-ups, especially those that involve making measurements, will allow you to demonstrate these skills. For example, an investigation of reaction rates of an enzyme at different temperatures might involve plotting a graph of product made against time and calculating the gradient of this graph to give the rate. You will need to: • include a graph, a table of results and a relevant diagram; • make certain that you explain why you made the calculations you did; • justify the way you presented your data.

8 Biological vocabulary

Some Latin and Greek roots of scientific terms and their meaning in English

a, an	*not, without*	ecto	*outside*
ab, abs	*away from*	encephalo	*brain*
abdomen, abdomini	*belly*	endo	*inside*
acro	*peak, summit*	entero	*gut, intestine*
adipi	*fat*	ephemero	*brief*
aero	*air*	erythro	*red*
allo, allelo	*different*	ex, e, ec	*out of*
ambi	*both*		
amphi	*both*	fer	*bearing*
amylo	*starch*	fili	*thread*
andro	*male*	flagelli	*lash, whip*
anneli	*small ring*	fove	*groove*
ante	*before*		
anthropo	*human*	galacto	*milk*
arachno	*spider*	gaster, gastro	*stomach*
arthro	*joint*	glutin	*blue*
asco	*sac*	glyco, glycy, gluco	*sweet, sugar*
astero, astro	*star*	gramini	*grass*
auri	*ear*	gymno	*naked*
auto	*self*		
auxo	*growth*	haemo, haem	*blood*
		halo	*salt*
bacilli	*small stick*	helix, helico	*coiled*
basidio	*column*	helmintho	*worm*
bi, bin	*two*	hemi	*half*
bio	*life*	hepato	*liver*
blasto	*bud, origin*	hetero	*different*
brachy	*short*	holo	*complete*
branchio	*gill*	hydro	*water*
bronchio	*throat, windpipe*	hygro	*moisture*
bucca	*mouth*	hyper	*excess*
		hypo	*below*
calyx, calyco	*cup*		
calyptro	*covering*	ichthyo	*fish*
cardio	*heart*	inter	*between*
cephalo	*head*	intra	*within*
chaeto	*bristle*	iris, irido	*rainbow*
chloro	*green*	iso	*the same*
coelo	*cavity*		
costa	*rib*	lacti	*milk*
cuti	*skin*	lamelli	*small place*
cyano	*blue*	latero	*slide*
cysto	*bladder*	lepido	*scale*
cyto	*cell*	leuco	*white, colourless*
		limno	*lake*
dactylo	*finger, digit*	lipo	*fat*
demi	*a half*	loci	*place*
dendro	*tree, branch*	luteo	*yellow*
denti	*tooth*	lyso	*loosening, breakdown*
dermo, dermato	*skin*		
di	*two*	macro	*large*
dictyo	*net*	malaco	*soft*
dino	*terrible*	mammi	*breast*
diplo	*double*	medulli	*pith*
dorso	*back*	mega	*large*

melano	*black*	proto	*first*
meso	*middle*	proximi	*near*
meta	*after*	pseudo	*false*
micro	*small*	pteris, pterido	*fern*
mito	*thread*	pulmo	*lung*
morpho	*form*		
multi	*many*	reni	*kidney*
myco, myceto	*fungus*	rhizo, rrhizo	*root*
myo, mys	*muscle*		
		sacci	*sac*
nano, nanno	*dwarf*	saccharo	*sugar*
necro	*dead, corpse*	sapro	*decaying*
nemo, nemato	*thread*	sarco	*flesh*
nephro	*kidney*	sauro	*lizard*
neuro	*nerve*	schizo	*split*
		scroti	*pouch*
oculo	*eye*	septi	*wall*
ocelli	*small eye*	somo, somato	*body*
oligo	*few*	spondylo	*vertebra*
operculi	*lid*	sporo	*spore, seed*
ops, opto	*sight, eye*	squamo	*scale*
ortho	*straight*	sterno	*chest*
osmo	*push*	suberi	*cork*
osteo	*bone*	sudori	*sweat*
ostraco	*shell*	sulci	*groove*
oto	*ear*		
		telo	*final*
paedo	*child*	terri	*land*
patelli	*dish*	thalamo	*chamber*
patho	*disease*	thermo	*heat*
pectori	*chest*	thorax, thoraco	*chest*
pedi	*foot, leg*	thrombo	*clot*
peri	*around*	trans	*across*
phago	*feeding*	tropo	*direction*
phello	*cork*	tropho	*feeding*
photo	*light*	turgo	*swollen*
phyco	*alga*	tympano	*drum*
phyllo	*leaf*		
plankto	*wandering*	uter, utro	*sack*
platy	*flat*		
pleio, pleo	*many*	villi	*shaggy*
pneumo	*lung*		
podo, pus	*foot, leg*	xantho	*yellow*
poecilo, poikilo	*various*	xero	*dry*
poly	*many*		
pre	*before*	zoo	*animal*
pro	*before*	zygo	*yoke, fusion*

9 Key figures from *New Understanding Biology for Advanced Level (4th edition)*

Part I   Levels of Organization

Chapter 2

Molecular organization

Fig. 2.1   Structure of various isomers of glucose and fructose
Fig. 2.2   Formation of maltose and sucrose
Fig. 2.3   Comparison of the properties and structures of amylose and amylopectin
Fig. 2.4   Structure of the cellulose molecule
Fig. 2.5   Formation of a triglyceride
Fig. 2.6   Structure of a phospholipid
Fig. 2.7   Structure of a range of amino acids
Fig. 2.8   Zwitterion formation in amino acids
Fig. 2.9   Formation of a dipeptide
Fig. 2.10   Formation of a polypeptide
Fig. 2.11   Types of bond in a polypeptide chain
Fig. 2.13   Structure of proteins

Chapter 3

Enzymes

Fig. 3.2   How enzymes lower the activation energy
Fig. 3.3   Mechanism of enzyme action
Fig. 3.8   Competitive inhibition
Fig. 3.9   Non-competitive inhibition

Chapter 4

Cellular organization

Fig. 4.1   Structure of the prokaryotic cell, e.g. a generalized bacterial cell
Fig. 4.2   A generalized animal cell with details of organelles
Fig. 4.3   A generalized plant cell with details of organelles
Fig. 4.4   The fluid-mosaic model of the cell membrane

Fig. 4.5   Structure of the chloroplasts
Fig. 4.7   Mitochondria
Fig. 4.8   Structure of a stalked granule
Fig. 4.9   Structure of rough endoplasmic reticulum
Fig. 4.10   The Golgi apparatus and its relationship to the nucleus, endoplasmic reticulum and lysosomes
Fig. 4.12   Structure of a ribosome
Fig. 4.13   Arrangement of alpha- and beta-tubulin within a microtubule
Fig. 4.18   Facilitated diffusion by carrier proteins
Fig. 4.21   Active transport

Chapter 5

Biodiversity

Fig. 5.1   Simplified diagram of tobacco mosaic virus
Fig. 5.2   Structure of a bacteriophage
Fig. 5.3   Life-cycle of a lytic (virulent) phage (e.g. T_2 phage)
Fig. 5.4   Life-cycle of a lysogenic (temperate) phage (e.g. λ phage)
Fig. 5.6   *Mucor*
Fig. 5.10   *Chlamydomonas*
Fig. 5.12   *Fucus*
Fig. 5.14   Life-cycle of a moss
Fig. 5.15   Life-cycle of a fern
Fig. 5.16   A comparison of hydroid and medusoid phases
Fig. 5.19   *Fasciola*
Fig. 5.21   *Ascaris* (female)
Fig. 5.23   *Lumbricus*
Fig. 5.26   Locust
Fig. 5.27   Land snail – gastropod

Part II   The Continuity of Life

Chapter 7

DNA and the genetic code

Fig. 7.2   Structure of a typical nucleotide
Fig. 7.3   Structure of section of polynucleotide, e.g. RNA
Fig. 7.4   Structure of molecules in a nucleotide
Fig. 7.5   Structure of transfer RNA

Fig. 7.6   Basic structure of DNA
Fig. 7.7   The DNA double helix structure
Fig. 7.9   The replication of DNA
Fig. 7.11   Transcription
Fig. 7.12   Activation
Fig. 7.13   (a) Translation (b) Polypeptide formation
Fig. 7.14   Use of plasmid vector in gene cloning

Part III   Energetics

Part VI   Microorganisms, Biotechnology, Health and Disease

2 Molecular organization

Straight chain arrangements

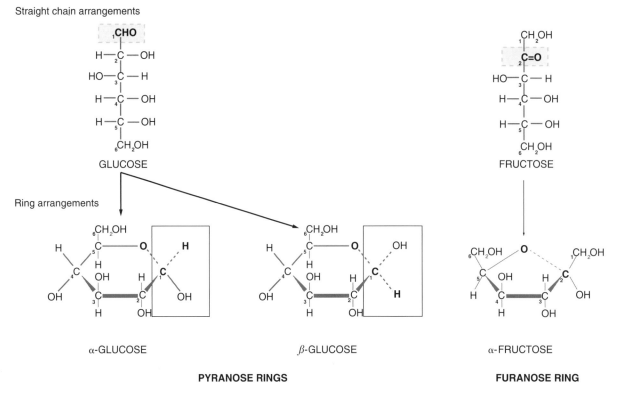

Ring arrangements

PYRANOSE RINGS   **FURANOSE RING**

Fig 2.1 Structure of various isomers of glucose and fructose

2 Molecular organization

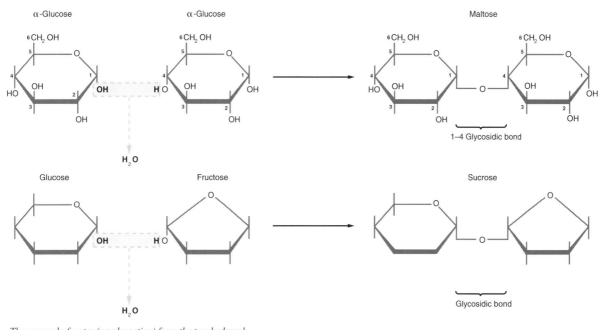

The removal of water (condensation) from the two hydroxyl groups (—OH) on carbons 1 and 4 of the respective glucose molecules forms a maltose molecule. Some carbon and hydrogen atoms have been omitted for simplicity.
Sucrose is formed by a condensation reaction between one glucose and one fructose molecule. The process shown is much simplified.

Fig 2.2 Formation of maltose and sucrose

2 Molecular organization

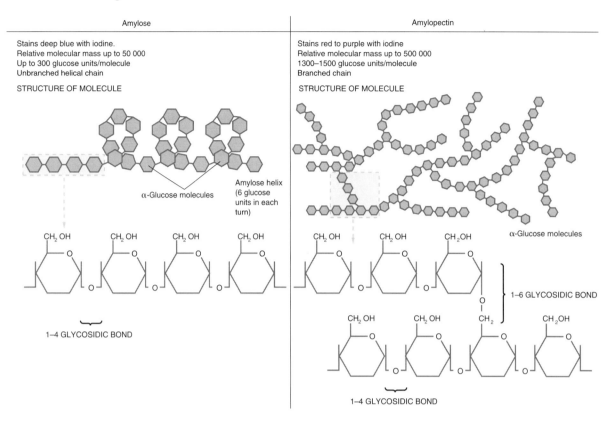

Amylose	Amylopectin
Stains deep blue with iodine.	Stains red to purple with iodine
Relative molecular mass up to 50 000	Relative molecular mass up to 500 000
Up to 300 glucose units/molecule	1300–1500 glucose units/molecule
Unbranched helical chain	Branched chain

STRUCTURE OF MOLECULE STRUCTURE OF MOLECULE

α-Glucose molecules

Amylose helix (6 glucose units in each turn)

α-Glucose molecules

CH₂OH

1–4 GLYCOSIDIC BOND

1–6 GLYCOSIDIC BOND

1–4 GLYCOSIDIC BOND

Fig 2.3 Comparison of the properties and structures of amylose and amylopectin

Being composed of β-glucose units, the chain, unlike that of starch, has adjacent glucose molecules rotated by 180°. This allows hydrogen bonds to be formed between the hydroxyl (—OH) groups on adjacent parallel chains which help to give cellulose its structural stability.

Simplified representation of the arrangement of glucose chains

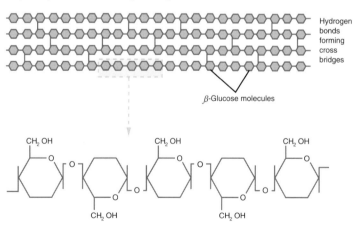

Hydrogen bonds forming cross bridges

β-Glucose molecules

Fig 2.4 Structure of the cellulose molecule

26

2 Molecular organization

The three triglycerides may all be the same, thereby forming a simple triglyceride, or they may be different in which case a mixed triglyceride is produced. In either case it is a condensation reaction.

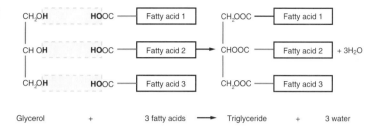

Fig 2.5 Formation of a triglyceride

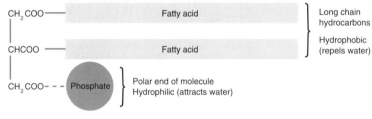

Fig 2.6 Structure of a phospholipid

2 Molecular organization

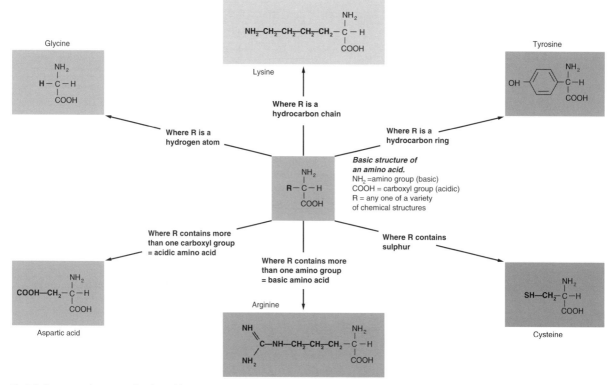

Fig 2.7 Structure of a range of amino acids

28

2 Molecular organization

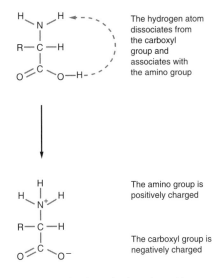

The hydrogen atom dissociates from the carboxyl group and associates with the amino group

The amino group is positively charged

The carboxyl group is negatively charged

Fig 2.8 Zwitterion formation in amino acids

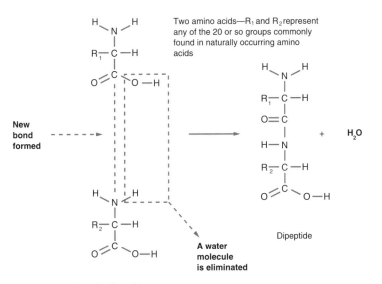

Two amino acids—R_1 and R_2 represent any of the 20 or so groups commonly found in naturally occurring amino acids

New bond formed

A water molecule is eliminated

Dipeptide

$+$ H_2O

Fig 2.9 Formation of a dipeptide

2 Molecular organization

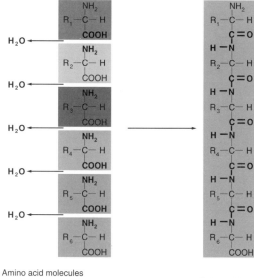

Amino acid molecules
(where R_1, R_2, R_3 etc. represent
any of the 20 or so groups
found in naturally occurring amino
acids)

Polypeptide
(part of)

Fig 2.10 Formation of a polypeptide

*A simplified representation of a polypeptide chain to show
three types of bonding responsible for shaping the chain. In
practice the polypeptide chains are longer, contain more of
these three types of bond and have a three-dimensional shape.*

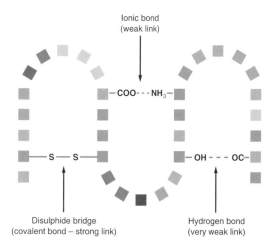

Ionic bond
(weak link)

$-COO---NH_3-$

$-S-S-$

$-OH---OC-$

Disulphide bridge
(covalent bond – strong link)

Hydrogen bond
(very weak link)

Fig 2.11 Types of bond in a polypeptide chain

2 Molecular organization

(a) The primary structure of a protein is the sequence of amino acids found in its polypeptide chains. This sequence determines its properties and shape. Following the elucidation of the amino acid sequence of the hormone insulin, by Frederick Sanger in 1954, the primary structure of many other proteins is now known.

	Lysine	Aspartic acid	Cysteine	Alanine	Tyrosine	Lysine	Glutamic acid	Valine	Glycine	

(b) The secondary structure is the shape which the polypeptide chain forms as a result of hydrogen bonding. This is most often a spiral known as the α-helix, although other configurations occur.

(c) The tertiary structure is due to the bending and twisting of the polypeptide helix into a compact structure. All three types of bond, disulphide, ionic and hydrogen, contribute to the maintenance of the tertiary structure.

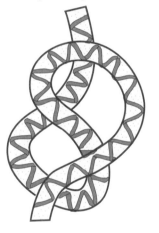

(d) The quaternary structure arises from the combination of a number of different polypeptide chains, and associated non-protein groups, into a large complex protein molecule.

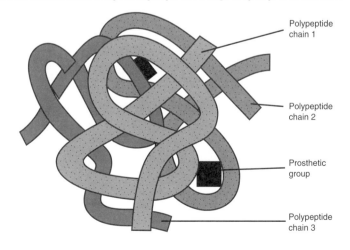

Polypeptide chain 1

Polypeptide chain 2

Prosthetic group

Polypeptide chain 3

Fig 2.13 Structure of proteins

3 Enzymes

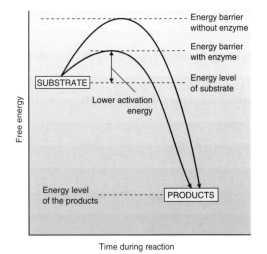

Fig 3.2 How enzymes lower the activation energy

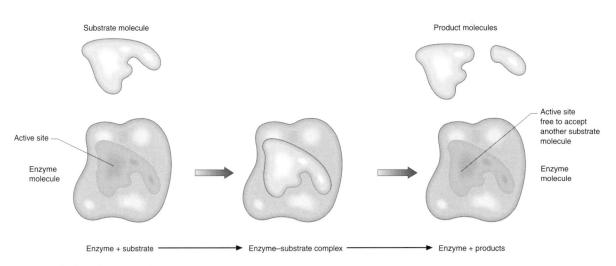

Fig 3.3 Mechanism of enzyme action

3 Enzymes

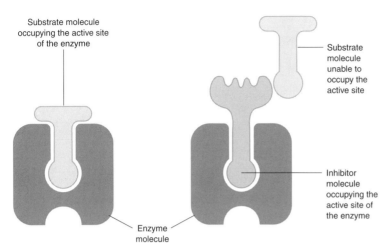

Fig 3.8 Competitive inhibition

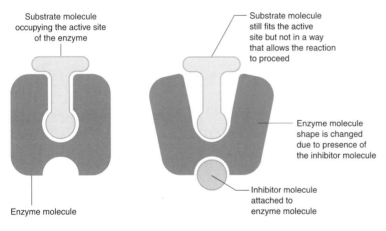

1. Inhibitor absent –
   The substrate attaches to the active site of the enzyme in the normal way. Reaction takes place as normal.
2. Inhibitor present –
   The inhibitor prevents the normal enzyme–substrate complex being formed. The reaction rate is reduced.

Fig 3.9 Non-competitive inhibition

4 Cellular organization

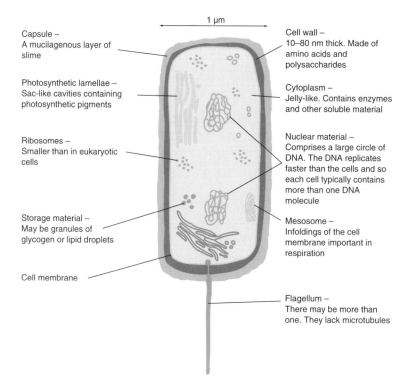

Capsule –
A mucilagenous layer of slime

Photosynthetic lamellae –
Sac-like cavities containing photosynthetic pigments

Ribosomes –
Smaller than in eukaryotic cells

Storage material –
May be granules of glycogen or lipid droplets

Cell membrane

1 μm

Cell wall –
10–80 nm thick. Made of amino acids and polysaccharides

Cytoplasm –
Jelly-like. Contains enzymes and other soluble material

Nuclear material –
Comprises a large circle of DNA. The DNA replicates faster than the cells and so each cell typically contains more than one DNA molecule

Mesosome –
Infoldings of the cell membrane important in respiration

Flagellum –
There may be more than one. They lack microtubules

Fig 4.1 Structure of the prokaryotic cell, e.g. a generalized bacterial cell

4 Cellular organization

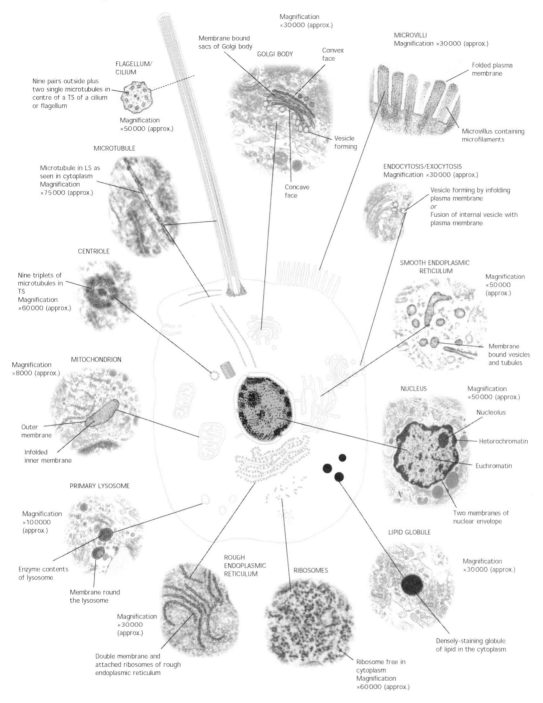

FLAGELLUM/
CILIUM

Nine pairs outside plus
two single microtubules in
centre of a TS of a cilium
or flagellum

Magnification
×50 000 (approx.)

MICROTUBULE

Microtubule in LS as
seen in cytoplasm
Magnification
×75 000 (approx.)

CENTRIOLE

Nine triplets of
microtubules in
TS
Magnification
×60 000 (approx.)

MITOCHONDRION

Magnification
×8 000 (approx.)

Outer
membrane

Infolded
inner membrane

PRIMARY LYSOSOME

Magnification
×100 000
(approx.)

Enzyme contents
of lysosome

Membrane round
the lysosome

Magnification
×30 000 (approx.)

Membrane bound
sacs of Golgi body

GOLGI BODY

Convex
face

Vesicle
forming

Concave
face

ROUGH
ENDOPLASMIC
RETICULUM

Magnification
×30 000
(approx.)

Double membrane and
attached ribosomes of rough
endoplasmic reticulum

RIBOSOMES

Ribosome free in
cytoplasm
Magnification
×60 000 (approx.)

MICROVILLI
Magnification ×30 000 (approx.)

Folded plasma
membrane

Microvillus containing
microfilaments

ENDOCYTOSIS/EXOCYTOSIS
Magnification ×30 000 (approx.)

Vesicle forming by infolding
plasma membrane
or
Fusion of internal vesicle with
plasma membrane

SMOOTH ENDOPLASMIC
RETICULUM

Magnification
×50 000
(approx.)

Membrane
bound vesicles
and tubules

NUCLEUS

Magnification
×50 000 (approx.)

Nucleolus

Heterochromatin

Euchromatin

Two membranes of
nuclear envelope

LIPID GLOBULE

Magnification
×30 000 (approx.)

Densely-staining globule
of lipid in the cytoplasm

Fig 4.2 A generalized animal cell with details of organelles

4 Cellular organization

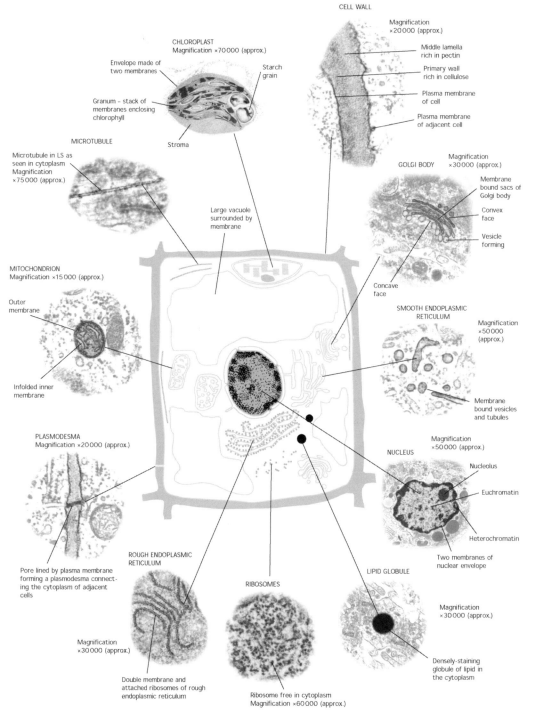

CELL WALL

Magnification ×20 000 (approx.)

Middle lamella rich in pectin

Primary wall rich in cellulose

Plasma membrane of cell

Plasma membrane of adjacent cell

CHLOROPLAST
Magnification ×70 000 (approx.)

Envelope made of two membranes

Starch grain

Granum – stack of membranes enclosing chlorophyll

Stroma

MICROTUBULE

Microtubule in LS as seen in cytoplasm
Magnification ×75 000 (approx.)

Large vacuole surrounded by membrane

GOLGI BODY

Magnification ×30 000 (approx.)

Membrane bound sacs of Golgi body

Convex face

Vesicle forming

Concave face

MITOCHONDRION
Magnification ×15 000 (approx.)

Outer membrane

Infolded inner membrane

SMOOTH ENDOPLASMIC RETICULUM

Magnification ×50 000 (approx.)

Membrane bound vesicles and tubules

PLASMODESMA
Magnification ×20 000 (approx.)

Magnification ×50 000 (approx.)

NUCLEUS

Nucleolus

Euchromatin

Heterochromatin

Two membranes of nuclear envelope

Pore lined by plasma membrane forming a plasmodesma connecting the cytoplasm of adjacent cells

ROUGH ENDOPLASMIC RETICULUM

RIBOSOMES

LIPID GLOBULE

Magnification ×30 000 (approx.)

Magnification ×30 000 (approx.)

Double membrane and attached ribosomes of rough endoplasmic reticulum

Ribosome free in cytoplasm
Magnification ×60 000 (approx.)

Densely-staining globule of lipid in the cytoplasm

Fig 4.3 A generalized plant cell with details of organelles

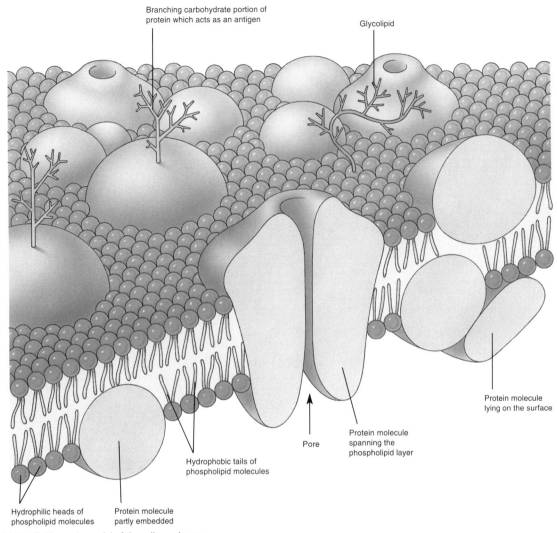

Branching carbohydrate portion of
protein which acts as an antigen

Glycolipid

Protein molecule
lying on the surface

Pore

Protein molecule
spanning the
phospholipid layer

Hydrophobic tails of
phospholipid molecules

Hydrophilic heads of
phospholipid molecules

Protein molecule
partly embedded

Fig 4.4 The fluid-mosaic model of the cell membrane

4 Cellular organization

(a)

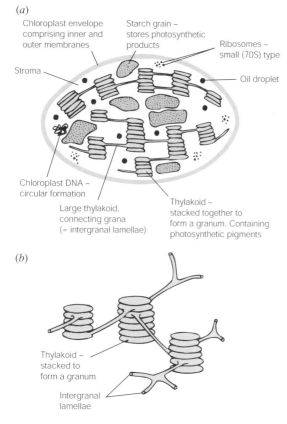

Chloroplast envelope comprising inner and outer membranes

Starch grain – stores photosynthetic products

Ribosomes – small (70S) type

Stroma

Oil droplet

Chloroplast DNA – circular formation

Large thylakoid, connecting grana (= intergranal lamellae)

Thylakoid – stacked together to form a granum. Containing photosynthetic pigments

(b)

Thylakoid – stacked to form a granum

Intergranal lamellae

Fig 4.5 Structure of the chloroplasts

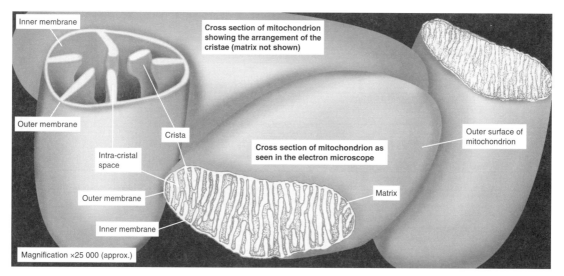

Inner membrane

Cross section of mitochondrion showing the arrangement of the cristae (matrix not shown)

Outer membrane

Crista

Outer surface of mitochondrion

Intra-cristal space

Cross section of mitochondrion as seen in the electron microscope

Outer membrane

Matrix

Inner membrane

Magnification ×25 000 (approx.)

Fig 4.7 Mitochondria

4 Cellular organization

Head piece

Stalk

Base piece

Fig 4.8 Structure of a stalked granule

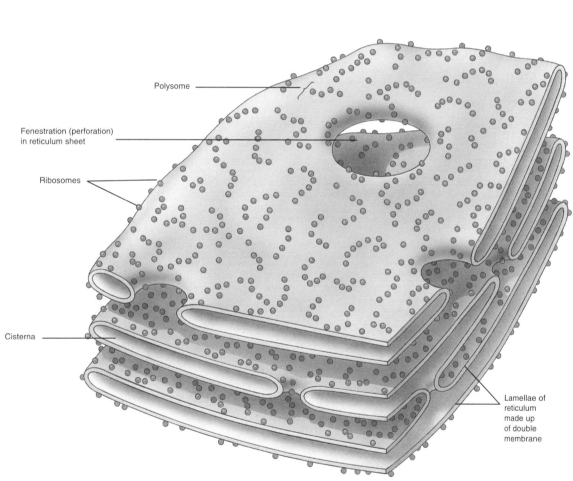

Polysome

Fenestration (perforation) in reticulum sheet

Ribosomes

Cisterna

Lamellae of reticulum made up of double membrane

Fig 4.9 Structure of rough endoplasmic reticulum

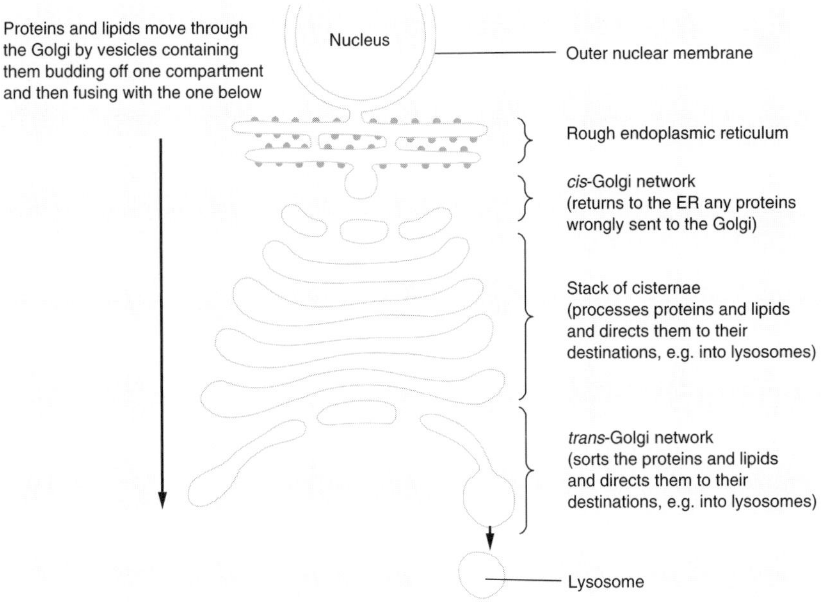

4 Cellular organization

Proteins and lipids move through the Golgi by vesicles containing them budding off one compartment and then fusing with the one below

Nucleus

Outer nuclear membrane

Rough endoplasmic reticulum

cis-Golgi network
(returns to the ER any proteins wrongly sent to the Golgi)

Stack of cisternae
(processes proteins and lipids and directs them to their destinations, e.g. into lysosomes)

trans-Golgi network
(sorts the proteins and lipids and directs them to their destinations, e.g. into lysosomes)

Lysosome

Fig 4.10 The Golgi apparatus and its relationship to the nucleus, endoplasmic reticulum and lysosomes

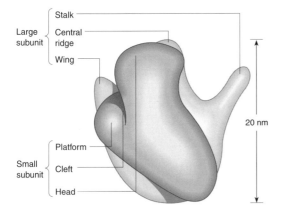

Stalk

Large subunit

Central ridge

Wing

20 nm

Platform

Small subunit

Cleft

Head

Fig 4.12 Structure of a ribosome

4 Cellular organization

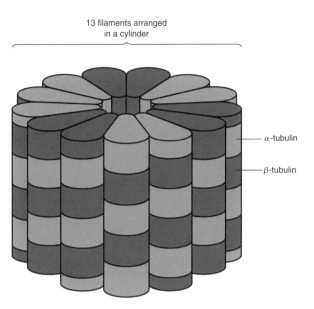

13 filaments arranged
in a cylinder

α-tubulin

β-tubulin

Fig 4.13 Arrangement of alpha- and beta-tubulin within a microtubule

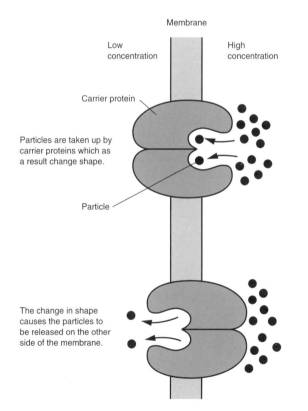

Membrane

Low
concentration

High
concentration

Carrier protein

Particles are taken up by
carrier proteins which as
a result change shape.

Particle

The change in shape
causes the particles to
be released on the other
side of the membrane.

Fig 4.18 Facilitated diffusion by carrier proteins

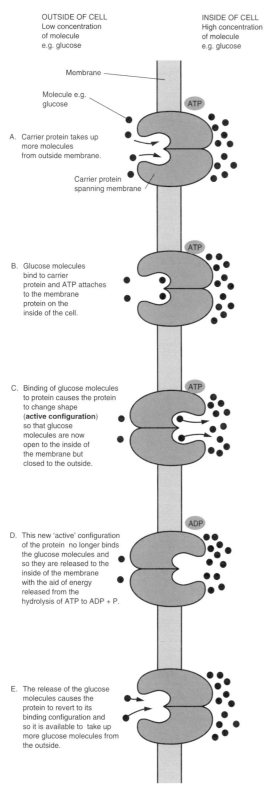

OUTSIDE OF CELL
Low concentration
of molecule
e.g. glucose

INSIDE OF CELL
High concentration
of molecule
e.g. glucose

Membrane

Molecule e.g.
glucose

ATP

A. Carrier protein takes up
more molecules
from outside membrane.

Carrier protein
spanning membrane

ATP

B. Glucose molecules
bind to carrier
protein and ATP attaches
to the membrane
protein on the
inside of the cell.

ATP

C. Binding of glucose molecules
to protein causes the protein
to change shape
(**active configuration**)
so that glucose
molecules are now
open to the inside of
the membrane but
closed to the outside.

ADP

D. This new 'active' configuration
of the protein no longer binds
the glucose molecules and
so they are released to the
inside of the membrane
with the aid of energy
released from the
hydrolysis of ATP to ADP + P.

E. The release of the glucose
molecules causes the
protein to revert to its
binding configuration and
so it is available to take up
more glucose molecules from
the outside.

Fig 4.21 Active transport

5 Biodiversity

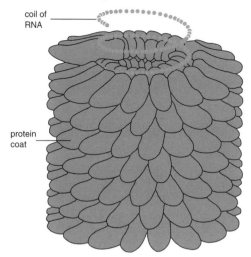

coil of RNA

protein coat

Fig 5.1 Simplified diagram of tobacco mosaic virus

5 Biodiversity

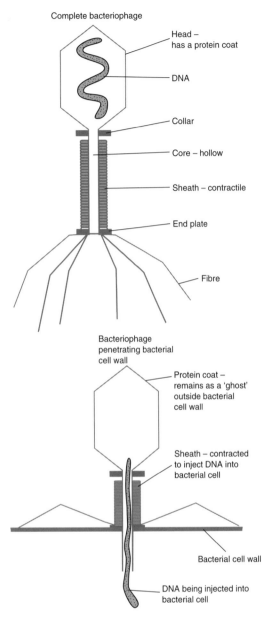

Complete bacteriophage

Head –
has a protein coat

DNA

Collar

Core – hollow

Sheath – contractile

End plate

Fibre

Bacteriophage
penetrating bacterial
cell wall

Protein coat –
remains as a 'ghost'
outside bacterial
cell wall

Sheath – contracted
to inject DNA into
bacterial cell

Bacterial cell wall

DNA being injected into
bacterial cell

Fig 5.2 Structure of a bacteriophage

5 Biodiversity

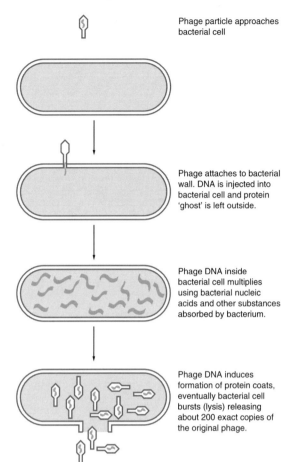

Phage particle approaches
bacterial cell

Phage attaches to bacterial
wall. DNA is injected into
bacterial cell and protein
'ghost' is left outside.

Phage DNA inside
bacterial cell multiplies
using bacterial nucleic
acids and other substances
absorbed by bacterium.

Phage DNA induces
formation of protein coats,
eventually bacterial cell
bursts (lysis) releasing
about 200 exact copies of
the original phage.

Fig 5.3 Life-cycle of a lytic (virulent) phage (e.g. T_2 phage)

5 Biodiversity

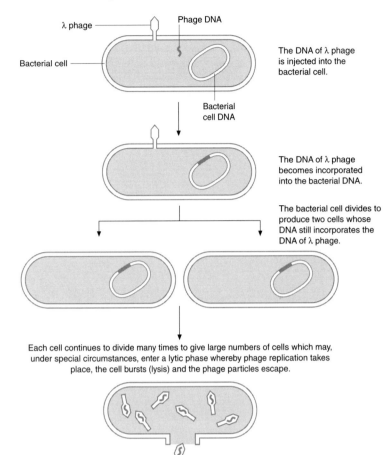

λ phage ——— Phage DNA

Bacterial cell ———

The DNA of λ phage is injected into the bacterial cell.

Bacterial cell DNA

The DNA of λ phage becomes incorporated into the bacterial DNA.

The bacterial cell divides to produce two cells whose DNA still incorporates the DNA of λ phage.

Each cell continues to divide many times to give large numbers of cells which may, under special circumstances, enter a lytic phase whereby phage replication takes place, the cell bursts (lysis) and the phage particles escape.

Fig 5.4 Life-cycle of a lysogenic (temperate) phage (e.g. λ phage)

5 Biodiversity

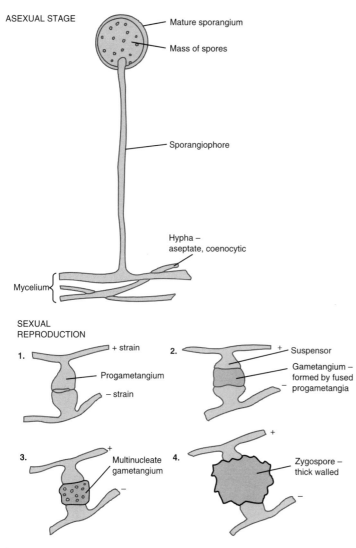

ASEXUAL STAGE

Mature sporangium

Mass of spores

Sporangiophore

Hypha –
aseptate, coenocytic

Mycelium

SEXUAL
REPRODUCTION

1. + strain

Progametangium

– strain

2. + Suspensor

Gametangium –
formed by fused
progametangia

3. + Multinucleate
gametangium

–

4. + Zygospore –
thick walled

–

Fig 5.6 *Mucor*

5 Biodiversity

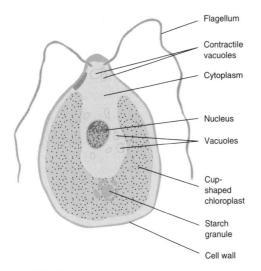

- Flagellum
- Contractile vacuoles
- Cytoplasm
- Nucleus
- Vacuoles
- Cup-shaped chloroplast
- Starch granule
- Cell wall

Fig 5.10 *Chlamydomonas*

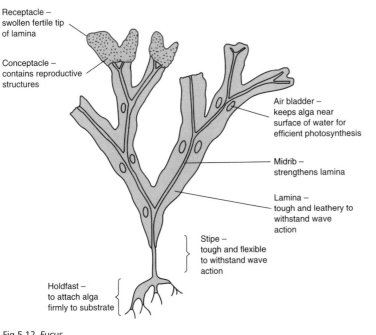

Receptacle – swollen fertile tip of lamina

Conceptacle – contains reproductive structures

Air bladder – keeps alga near surface of water for efficient photosynthesis

Midrib – strengthens lamina

Lamina – tough and leathery to withstand wave action

Stipe – tough and flexible to withstand wave action

Holdfast – to attach alga firmly to substrate

Fig 5.12 *Fucus*

5 Biodiversity

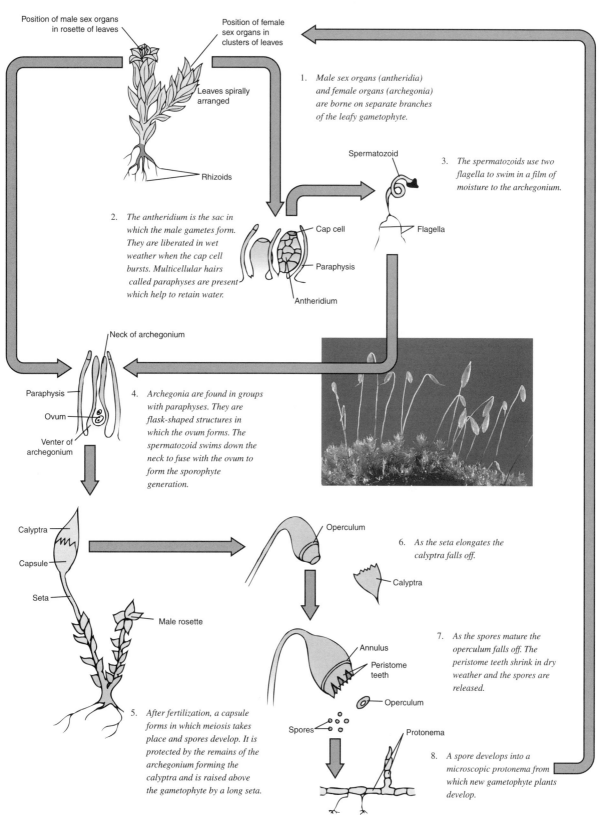

Position of male sex organs in rosette of leaves

Position of female sex organs in clusters of leaves

Leaves spirally arranged

Rhizoids

1. *Male sex organs (antheridia) and female organs (archegonia) are borne on separate branches of the leafy gametophyte.*

Spermatozoid

3. *The spermatozoids use two flagella to swim in a film of moisture to the archegonium.*

Flagella

Cap cell

Paraphysis

Antheridium

2. *The antheridium is the sac in which the male gametes form. They are liberated in wet weather when the cap cell bursts. Multicellular hairs called paraphyses are present which help to retain water.*

Neck of archegonium

Paraphysis

Ovum

Venter of archegonium

4. *Archegonia are found in groups with paraphyses. They are flask-shaped structures in which the ovum forms. The spermatozoid swims down the neck to fuse with the ovum to form the sporophyte generation.*

Calyptra

Capsule

Seta

Male rosette

Operculum

6. *As the seta elongates the calyptra falls off.*

Calyptra

Annulus

Peristome teeth

Operculum

Spores

Protonema

7. *As the spores mature the operculum falls off. The peristome teeth shrink in dry weather and the spores are released.*

5. *After fertilization, a capsule forms in which meiosis takes place and spores develop. It is protected by the remains of the archegonium forming the calyptra and is raised above the gametophyte by a long seta.*

8. *A spore develops into a microscopic protonema from which new gametophyte plants develop.*

Fig 5.14 Life-cycle of a moss

5 Biodiversity

1. *Dominant sporophyte is diploid. It consists of a number of fronds growing from an underground rhizome. It can photosynthesize, and the roots absorb water from the soil.*

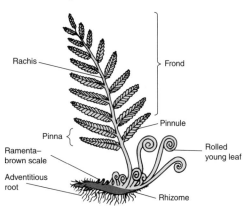

2. *In late summer, brown sori are visible on the underside of each pinnule. Each sorus is made up of a number of sporangia covered by the indusium.*

3. *The sporangia are found in groups. Within each sporangium, meiosis occurs to produce haploid spores.*

VS Pinnule

Pinnule
Indusium
Sporangium
Placenta

4. *As the spores mature the indusium falls off. The exposed sporangia dry out. The uneven thickening in the annulus sets up tension in the stomium whose cells rupture suddenly to release the spores.*

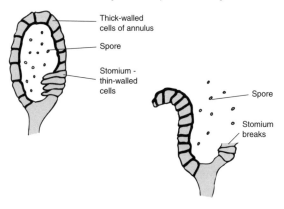

Thick-walled cells of annulus
Spore
Stomium - thin-walled cells
Spore
Stomium breaks

5. *The spores germinate in moist conditions and develop into a tiny plate of cells called the prothallus. This is the haploid gametophyte stage bearing archegonia and antheridia. It has rhizoids for anchorage and it is photosynthetic.*

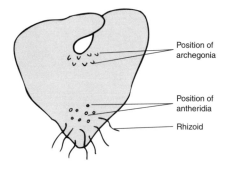

Position of archegonia
Position of antheridia
Rhizoid

6. *Antheridia develop on the under-surface of the prothallus near the rhizoids. Within them, spiral, multiflagellate spermatozoids develop.*

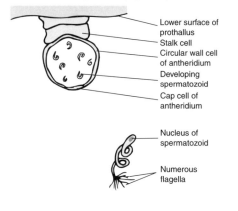

Lower surface of prothallus
Stalk cell
Circular wall cell of antheridium
Developing spermatozoid
Cap cell of antheridium

Nucleus of spermatozoid
Numerous flagella

7. *The archegonia develop after the antheridia, so self-fertilization is rare. The venter of the archegonium is embedded in the prothallus and the neck is short.*

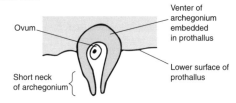

Venter of archegonium embedded in prothallus
Ovum
Lower surface of prothallus
Short neck of archegonium

8. *Following fertilization, the young diploid sporophyte plant grows to become an independent plant. At first it has a foot which absorbs nutrients from the prothallus.*

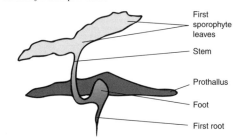

First sporophyte leaves
Stem
Prothallus
Foot
First root

Fig 5.15 Life-cycle of a fern

5 Biodiversity

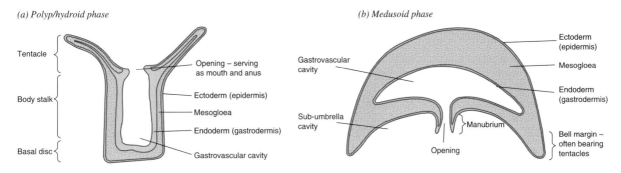

(a) Polyp/hydroid phase

Tentacle

Body stalk

Basal disc

Opening – serving as mouth and anus

Ectoderm (epidermis)

Mesogloea

Endoderm (gastrodermis)

Gastrovascular cavity

(b) Medusoid phase

Gastrovascular cavity

Sub-umbrella cavity

Opening

Manubrium

Ectoderm (epidermis)

Mesogloea

Endoderm (gastrodermis)

Bell margin – often bearing tentacles

Fig 5.16 A comparison of hydroid and medusoid phases

The excretory and reproductive systems are very diffuse, and cover the digestive system.

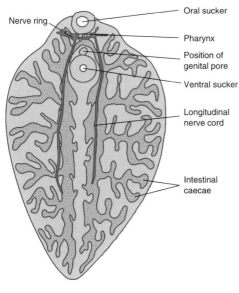

Nerve ring

Oral sucker

Pharynx

Position of genital pore

Ventral sucker

Longitudinal nerve cord

Intestinal caecae

Fig 5.19 *Fasciola*

5 Biodiversity

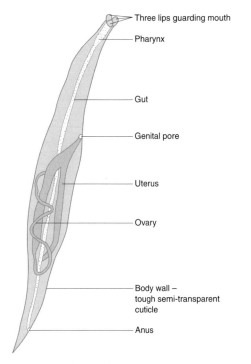

- Three lips guarding mouth
- Pharynx
- Gut
- Genital pore
- Uterus
- Ovary
- Body wall – tough semi-transparent cuticle
- Anus

Fig 5.21 *Ascaris* (female)

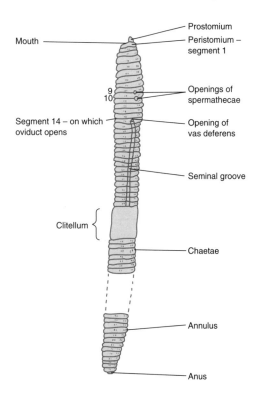

Mouth

- Prostomium
- Peristomium – segment 1
- Openings of spermathecae
- Opening of vas deferens
- Seminal groove
- Chaetae
- Annulus
- Anus

9
10

Segment 14 – on which oviduct opens

Clitellum

Fig 5.23 *Lumbricus*

5 Biodiversity

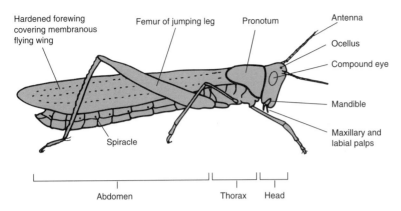

Hardened forewing covering membranous flying wing

Femur of jumping leg

Pronotum

Antenna

Ocellus

Compound eye

Mandible

Maxillary and labial palps

Spiracle

Abdomen

Thorax

Head

Fig 5.26 Locust

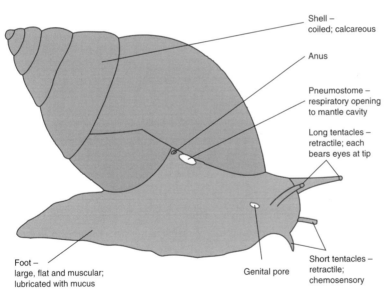

Shell – coiled; calcareous

Anus

Pneumostome – respiratory opening to mantle cavity

Long tentacles – retractile; each bears eyes at tip

Short tentacles – retractile; chemosensory

Genital pore

Foot – large, flat and muscular; lubricated with mucus

Fig 5.27 Land snail – gastropod

Adenosine monophosphate (adenylic acid)

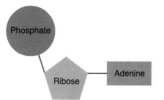

Fig 7.2 Structure of a typical nucleotide

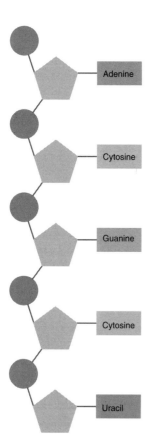

Fig 7.3 Structure of section of polynucleotide, e.g. RNA

7 DNA and the genetic code

NAME OF MOLECULE	CHEMICAL STRUCTURE	REPRESENTATIVE SHAPE
Phosphate	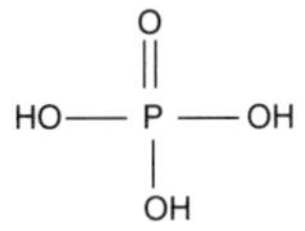	
Ribose	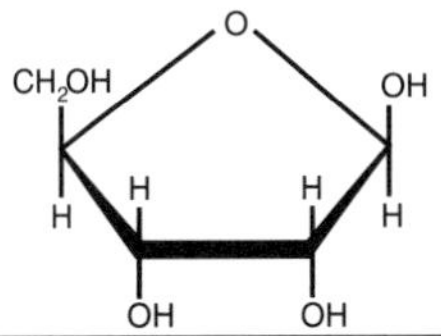	
Deoxyribose	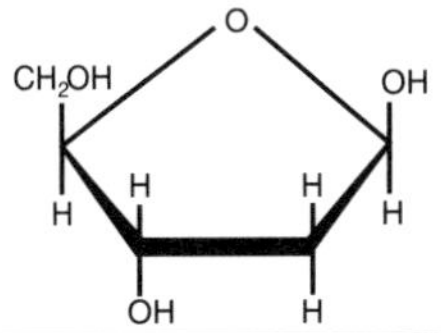	
Adenine (a purine)	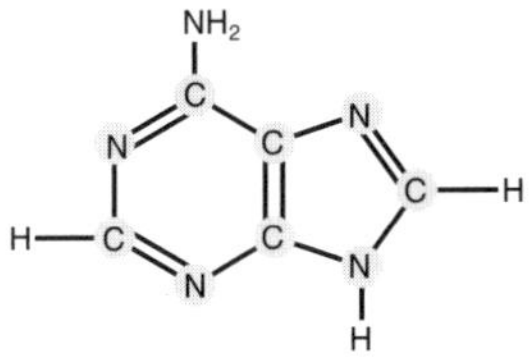	 Adenine
Guanine (a purine)	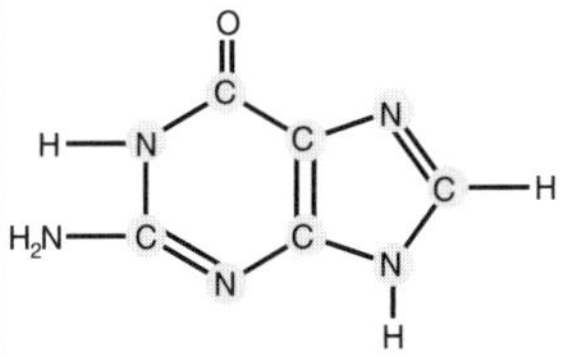	Guanine
Cytosine (a pyrimidine)	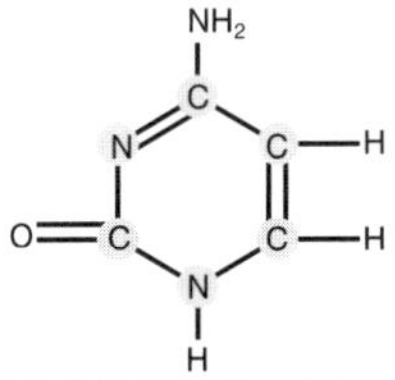	Cytosine
Thymine (a pyrimidine)	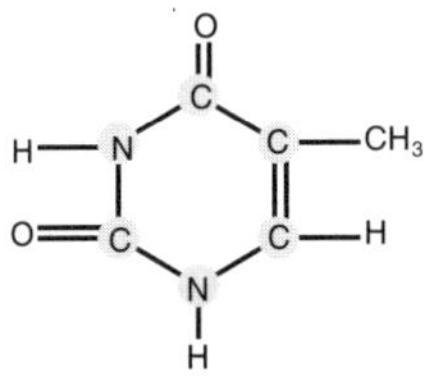	Thymine
Uracil (a pyrimidine)	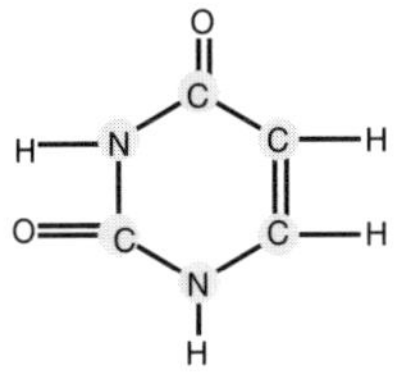	Uracil

Fig 7.4 Structure of molecules in a nucleotide

7 DNA and the genetic code

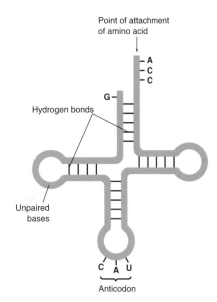

Fig 7.5 Structure of transfer RNA

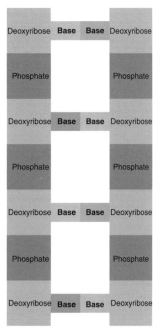

Simplified ladder

DNA structure may be likened to a ladder where alternating phosphate and deoxyribose molecules make up the 'uprights' and pairs of organic bases comprise the 'rungs'.

Fig 7.6 Basic structure of DNA

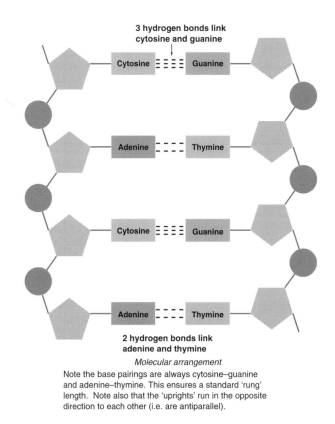

Molecular arrangement

Note the base pairings are always cytosine–guanine and adenine–thymine. This ensures a standard 'rung' length. Note also that the 'uprights' run in the opposite direction to each other (i.e. are antiparallel).

7 DNA and the genetic code

The uprights are composed of deoxyribose–phosphate molecules, the rungs of pairs of bases.

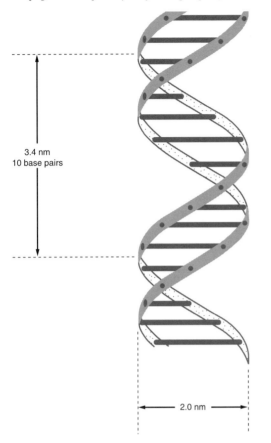

3.4 nm
10 base pairs

2.0 nm

Fig 7.7 The DNA double helix structure

7 DNA and the genetic code

1. *A representative portion of DNA, which is about to undergo replication, is shown.*

2. *DNA polymerase causes the two strands of the DNA to separate.*

3. *The DNA polymerase completes the splitting of the strand. Meanwhile free nucleotides are attracted to their complementary bases.*

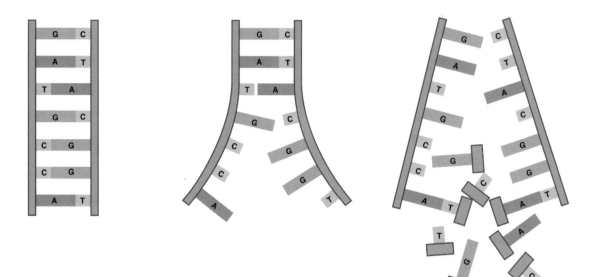

4. *Once the nucleotides are lined up they join together (bottom 3 nucleotides). The remaining unpaired bases continue to attract their complementary nucleotides.*

5. *Finally all the nucleotides are joined to form a complete polynucleotide chain. In this way two identical strands of DNA are formed. As each strand retains half of the original DNA material, this method of replication is called the semi-conservative method.*

Fig 7.9 The replication of DNA

7 DNA and the genetic code

A portion of DNA, called a cistron, unwinds. One strand acts as a template for the formation of mRNA

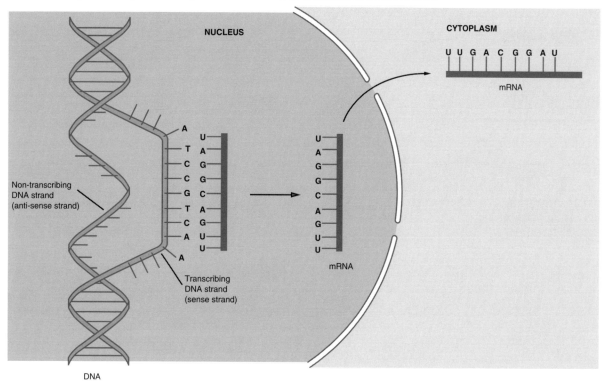

Fig 7.11 Transcription

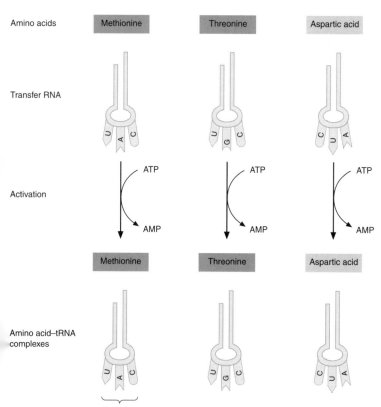

Fig 7.12 Activation

7 DNA and the genetic code

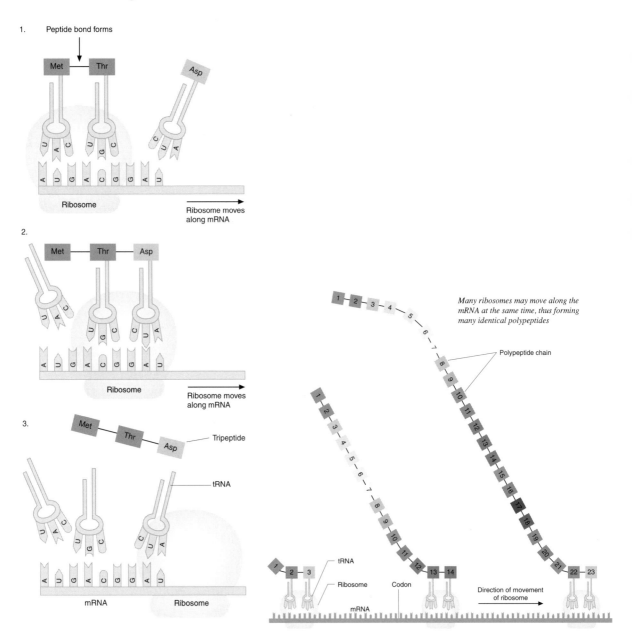

Fig 7.13 (a) Translation

(b) Polypeptide formation

7 DNA and the genetic code

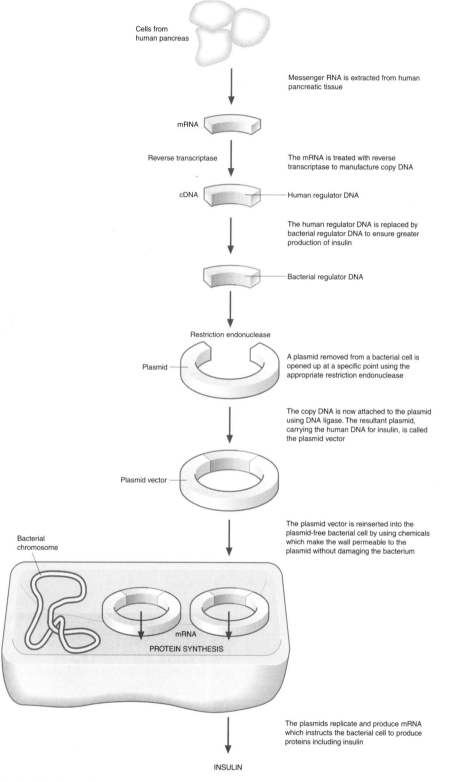

Cells from human pancreas

Messenger RNA is extracted from human pancreatic tissue

mRNA

Reverse transcriptase

The mRNA is treated with reverse transcriptase to manufacture copy DNA

cDNA — Human regulator DNA

The human regulator DNA is replaced by bacterial regulator DNA to ensure greater production of insulin

Bacterial regulator DNA

Restriction endonuclease

Plasmid

A plasmid removed from a bacterial cell is opened up at a specific point using the appropriate restriction endonuclease

The copy DNA is now attached to the plasmid using DNA ligase. The resultant plasmid, carrying the human DNA for insulin, is called the plasmid vector

Plasmid vector

The plasmid vector is reinserted into the plasmid-free bacterial cell by using chemicals which make the wall permeable to the plasmid without damaging the bacterium

Bacterial chromosome

mRNA

PROTEIN SYNTHESIS

The plasmids replicate and produce mRNA which instructs the bacterial cell to produce proteins including insulin

INSULIN

Fig 7.14 Use of plasmid vector in gene cloning

8 Cell division

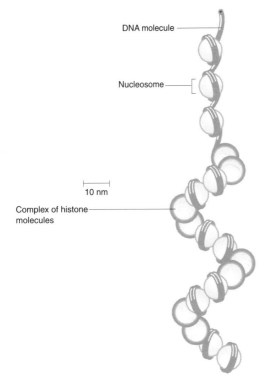

DNA molecule

Nucleosome

10 nm

Complex of histone molecules

Fig 8.2 Detailed structure of a chromosome

Nuclear division, or mitosis, typically occupies 5–10% of the total cycle. The cycle may take as little as 20 minutes in a bacterial cell, although it typically takes 8–24 hours.

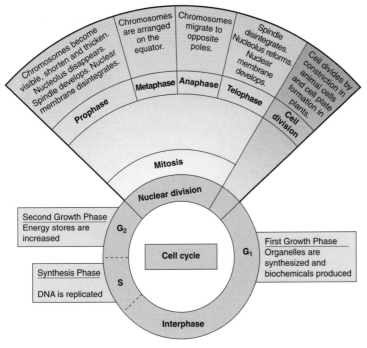

Chromosomes become visible, shorten and thicken. Nucleolus disappears. Spindle develops. Nuclear membrane disintegrates.

Prophase

Chromosomes are arranged on the equator.

Metaphase

Chromosomes migrate to opposite poles.

Anaphase

Spindle disintegrates. Nucleolus reforms. Nuclear membrane develops.

Telophase

Cell divides by constriction in animal cells and cell plate formation in plants.

Cell division

Mitosis

Nuclear division

Second Growth Phase Energy stores are increased

G_2

Cell cycle

G_1

First Growth Phase Organelles are synthesized and biochemicals produced

Synthesis Phase

S

DNA is replicated

Interphase

Fig 8.3 The cell cycle

8 Cell division

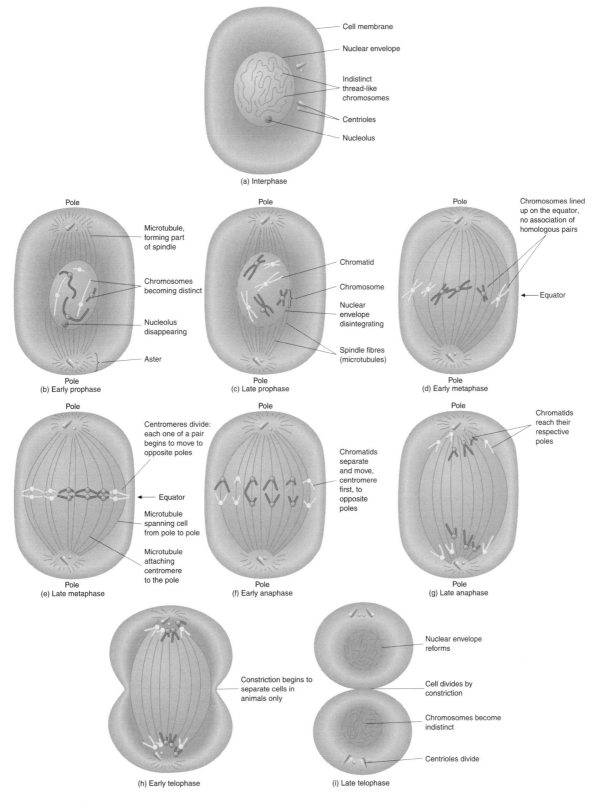

Fig 8.4 Stages of mitosis

8 Cell division

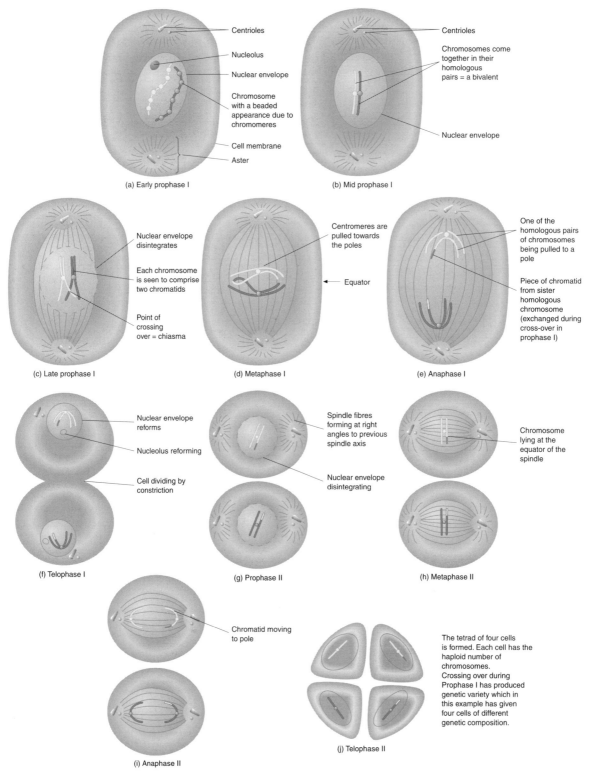

Centrioles

Nucleolus

Nuclear envelope

Chromosome with a beaded appearance due to chromomeres

Cell membrane

Aster

(a) Early prophase I

Centrioles

Chromosomes come together in their homologous pairs = a bivalent

Nuclear envelope

(b) Mid prophase I

Nuclear envelope disintegrates

Each chromosome is seen to comprise two chromatids

Point of crossing over = chiasma

(c) Late prophase I

Centromeres are pulled towards the poles

Equator

(d) Metaphase I

One of the homologous pairs of chromosomes being pulled to a pole

Piece of chromatid from sister homologous chromosome (exchanged during cross-over in prophase I)

(e) Anaphase I

Nuclear envelope reforms

Nucleolus reforming

Cell dividing by constriction

(f) Telophase I

Spindle fibres forming at right angles to previous spindle axis

Nuclear envelope disintegrating

(g) Prophase II

Chromosome lying at the equator of the spindle

(h) Metaphase II

Chromatid moving to pole

(i) Anaphase II

The tetrad of four cells is formed. Each cell has the haploid number of chromosomes. Crossing over during Prophase I has produced genetic variety which in this example has given four cells of different genetic composition.

(j) Telophase II

Fig 8.6 Stages of meiosis (only one pair of chromosomes shown)

9 Heredity and genetics

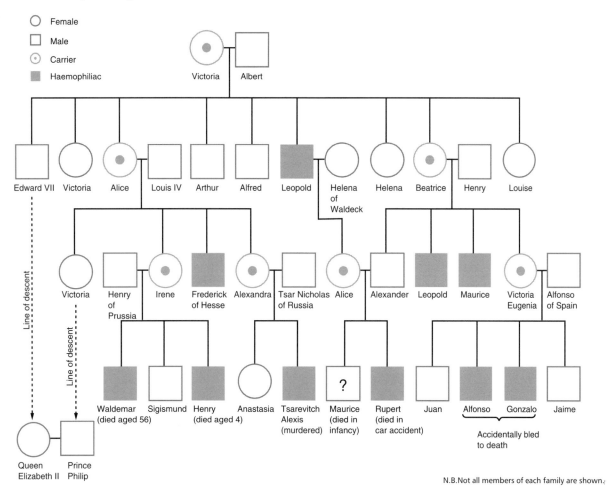

○ Female
□ Male
⊙ Carrier
▪ Haemophiliac

Victoria   Albert

Edward VII | Victoria | Alice | Louis IV | Arthur | Alfred | Leopold | Helena of Waldeck | Helena | Beatrice | Henry | Louise

Line of descent

Victoria | Henry of Prussia | Irene | Frederick of Hesse | Alexandra | Tsar Nicholas of Russia | Alice | Alexander | Leopold | Maurice | Victoria Eugenia | Alfonso of Spain

Line of descent

Waldemar (died aged 56) | Sigismund | Henry (died aged 4) | Anastasia | Tsarevitch Alexis (murdered) | Maurice (died in infancy) | Rupert (died in car accident) | Juan | Alfonso | Gonzalo | Jaime

? = Maurice (died in infancy)

Accidentally bled to death

Queen Elizabeth II | Prince Philip

N.B. Not all members of each family are shown.

Fig 9.3 Transmission of haemophilia from Queen Victoria

10 Genetic change and variation

1. *The DNA molecule which codes for the beta amino acid chain in haemoglobin has a mutation whereby the base adenine replaces thymine.*

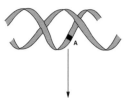

2. *The mRNA produced has the triplet codon GUA (for amino acid valine) rather than GAA (for amino acid glutamic acid).*

3. *The beta amino acid chain produced has one glutamic acid molecule replaced by a valine molecule.*

4. *The haemoglobin molecule containing the abnormal beta chains forms abnormal long fibres when the oxygen level of the blood is low. This haemoglobin is called haemoglobin-S.*

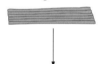

5. *Haemoglobin-S causes the shape of the red blood cell to become crescent (sickle) shaped.*

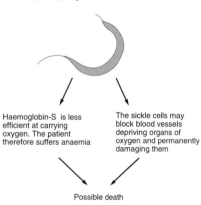

Haemoglobin-S is less efficient at carrying oxygen. The patient therefore suffers anaemia

The sickle cells may block blood vessels depriving organs of oxygen and permanently damaging them

Possible death

Fig 10.9 Sequence of events whereby a gene mutation causes sickle-cell anaemia

10 Genetic change and variation

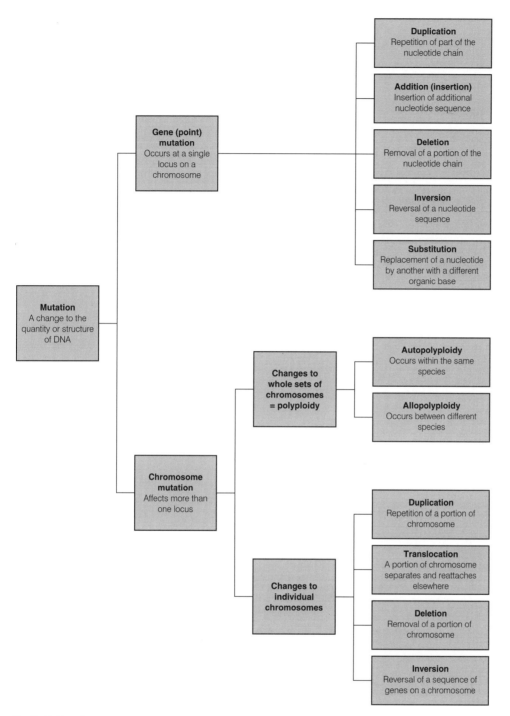

Fig 10.11 Mutations summary

11 Evolution

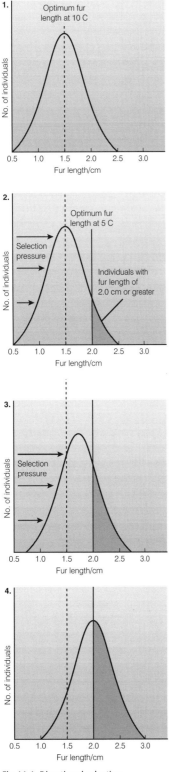

Fig 11.1 Directional selection

In a population of a particular mammal, fur length shows continuous variation.

1. When the average environmental temperature is 10°C, the optimum fur length is 1.5 cm. This then represents the mean fur length of the population.

2. A few individuals in the population already have a fur length of 2.0 cm or greater. If the average environmental temperature falls to 5°C, these individuals are better insulated and so are more likely to survive to breed. There is a selection pressure favouring individuals with longer fur.

3. The selection pressure causes a shift in the mean fur length towards longer fur over a number of generations. The selection pressure continues.

4. Over further generations the shift in the mean fur length continues until it reaches 2.0 cm – the optimum length for the prevailing average environmental temperature of 5°C. The selection pressure now ceases.

11 Evolution

1. *Initially there is a wide range of fur length about the mean of 1.5 cm. The fur lengths of less than 1.0 cm or greater than 2.0 cm in individuals are maintained by rapid breeding in years when the average temperature is much warmer or colder than normal.*

2. *When the average environmental temperature is consistently around 10°C with little annual variation, individuals with very long or very short hair are eliminated from the population over a number of generations.*

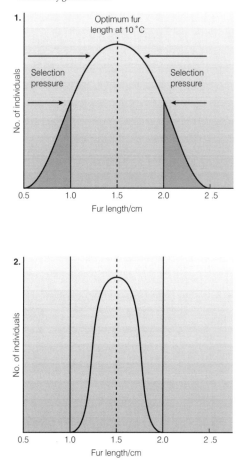

Fig 11.2 Stabilizing selection

11 Evolution

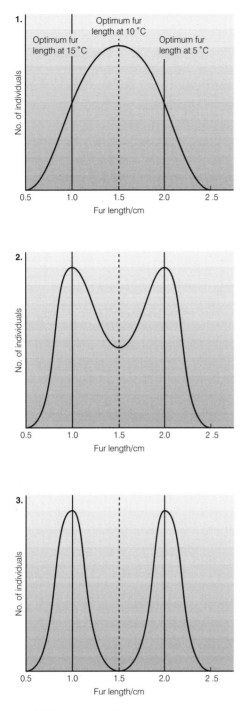

1.

Optimum fur length at 10°C

Optimum fur length at 15°C

Optimum fur length at 5°C

No. of individuals

0.5   1.0   1.5   2.0   2.5

Fur length/cm

2.

No. of individuals

0.5   1.0   1.5   2.0   2.5

Fur length/cm

3.

No. of individuals

0.5   1.0   1.5   2.0   2.5

Fur length/cm

Fig 11.3 Disruptive selection

1. *When there is a wide range of temperatures throughout the year, there is continuous variation in fur length around a mean of 1.5 cm.*

2. *Where the summer temperature is static around 15°C and the winter temperature is static around 5°C, individuals with two distinct fur lengths predominate: 1.0 cm types which are active in summer and 2.0 cm types which are active in winter.*

3. *After many generations two distinct sub-populations are formed.*

11 Evolution

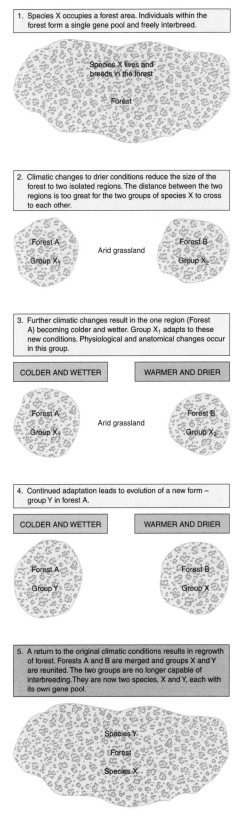

1. Species X occupies a forest area. Individuals within the forest form a single gene pool and freely interbreed.

Species X lives and breeds in the forest

Forest

2. Climatic changes to drier conditions reduce the size of the forest to two isolated regions. The distance between the two regions is too great for the two groups of species X to cross to each other.

Forest A

Group X_1

Arid grassland

Forest B

Group X_2

3. Further climatic changes result in the one region (Forest A) becoming colder and wetter. Group X_1 adapts to these new conditions. Physiological and anatomical changes occur in this group.

COLDER AND WETTER

WARMER AND DRIER

Forest A

Group X_1

Arid grassland

Forest B

Group X_2

4. Continued adaptation leads to evolution of a new form – group Y in forest A.

COLDER AND WETTER

WARMER AND DRIER

Forest A

Group Y

Forest B

Group X

5. A return to the original climatic conditions results in regrowth of forest. Forests A and B are merged and groups X and Y are reunited. The two groups are no longer capable of interbreeding. They are now two species, X and Y, each with its own gene pool.

Species Y

Forest

Species X

Fig 11.4 Speciation due to geographical isolation

12 Reproduction, development and growth

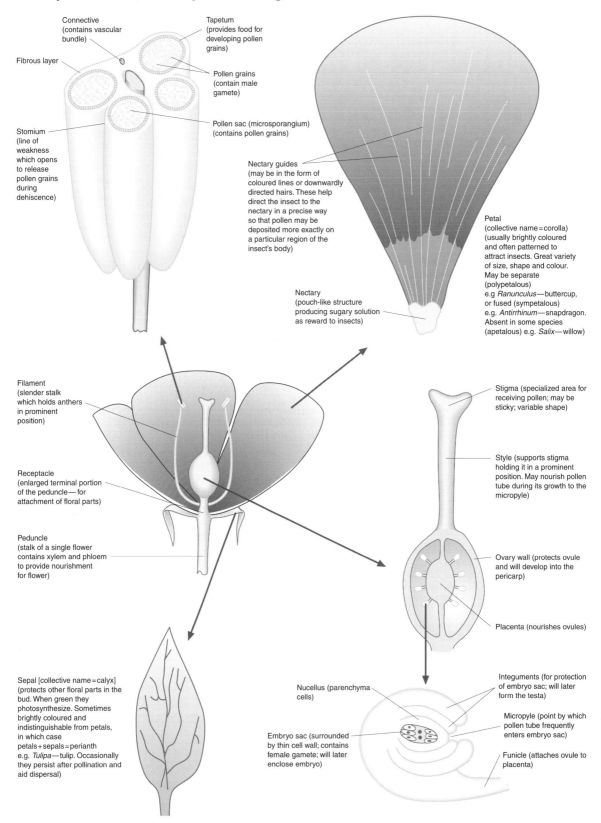

Connective (contains vascular bundle)

Fibrous layer

Tapetum (provides food for developing pollen grains)

Pollen grains (contain male gamete)

Stomium (line of weakness which opens to release pollen grains during dehiscence)

Pollen sac (microsporangium) (contains pollen grains)

Nectary guides (may be in the form of coloured lines or downwardly directed hairs. These help direct the insect to the nectary in a precise way so that pollen may be deposited more exactly on a particular region of the insect's body)

Petal (collective name = corolla) (usually brightly coloured and often patterned to attract insects. Great variety of size, shape and colour. May be separate (polypetalous) e.g *Ranunculus*—buttercup, or fused (sympetalous) e.g. *Antirrhinum*—snapdragon. Absent in some species (apetalous) e.g. *Salix*—willow)

Nectary (pouch-like structure producing sugary solution as reward to insects)

Filament (slender stalk which holds anthers in prominent position)

Receptacle (enlarged terminal portion of the peduncle—for attachment of floral parts)

Peduncle (stalk of a single flower contains xylem and phloem to provide nourishment for flower)

Stigma (specialized area for receiving pollen; may be sticky; variable shape)

Style (supports stigma holding it in a prominent position. May nourish pollen tube during its growth to the micropyle)

Ovary wall (protects ovule and will develop into the pericarp)

Placenta (nourishes ovules)

Sepal [collective name = calyx] (protects other floral parts in the bud. When green they photosynthesize. Sometimes brightly coloured and indistinguishable from petals, in which case petals + sepals = perianth e.g. *Tulipa*—tulip. Occasionally they persist after pollination and aid dispersal)

Nucellus (parenchyma cells)

Embryo sac (surrounded by thin cell wall; contains female gamete; will later enclose embryo)

Integuments (for protection of embryo sac; will later form the testa)

Micropyle (point by which pollen tube frequently enters embryo sac)

Funicle (attaches ovule to placenta)

Fig 12.3 Floral structure

12 Reproduction, development and growth

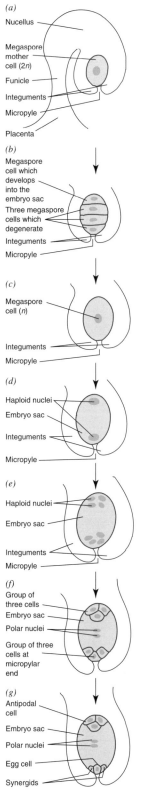

(a)
Nucellus
Megaspore mother cell (2n)
Funicle
Integuments
Micropyle
Placenta

(b)
Megaspore cell which develops into the embryo sac
Three megaspore cells which degenerate
Integuments
Micropyle

(c)
Megaspore cell (n)
Integuments
Micropyle

(d)
Haploid nuclei
Embryo sac
Integuments
Micropyle

(e)
Haploid nuclei
Embryo sac
Integuments
Micropyle

(f)
Group of three cells
Embryo sac
Polar nuclei
Group of three cells at micropylar end

(g)
Antipodal cell
Embryo sac
Polar nuclei
Egg cell
Synergids

(a) *The ovule consists of a mass of cells called the* **nucellus** *which is carried on a short stalk called the* **funicle.** *The nucellus is completely surrounded by two protective* **integuments** *except for a narrow channel at the tip called the* **micropyle.** *One cell of the nucellus becomes larger and more conspicuous than the rest. This is the* **embryo sac mother cell.**

(b) *The embryo sac mother cell divides meiotically to give four haploid* **megaspore cells.**

(c) *The three cells nearest the micropyle degenerate while the remaining one enlarges to form the* **embryo sac.**

(d) *The embryo sac nucleus divides by mitosis and the resultant nuclei migrate to opposite poles.*

(e) *Each nucleus undergoes two mitotic divisions to give a group of four haploid nuclei at each pole.*

(f) *One nucleus from each polar group moves to the centre of the embryo sac. These are the polar nuclei. The remaining nuclei develop cytoplasm around them and become separated by cell walls, leaving two groups of three cells at each pole.*

(g) *The three cells at the opposite end to the micropyle are called* **antipodal cells** *and play no further role in the process. Of the three cells at the micropyle end, one, the* **egg cell** *remains, the other two, the* **synergids,** *degenerate.*

Fig 12.4 Structure and development of the ovule

12 Reproduction, development and growth

The anthers comprise pollen sacs (usually four) which contain a mass of diploid pollen mother cells.

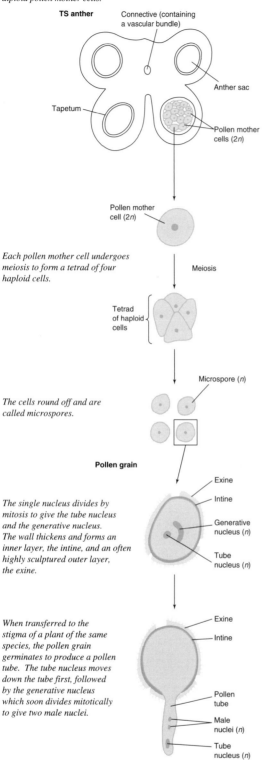

TS anther

Connective (containing a vascular bundle)

Anther sac

Tapetum

Pollen mother cells (2*n*)

Pollen mother cell (2*n*)

Each pollen mother cell undergoes meiosis to form a tetrad of four haploid cells.

Meiosis

Tetrad of haploid cells

Microspore (*n*)

The cells round off and are called microspores.

Pollen grain

The single nucleus divides by mitosis to give the tube nucleus and the generative nucleus. The wall thickens and forms an inner layer, the intine, and an often highly sculptured outer layer, the exine.

Exine

Intine

Generative nucleus (*n*)

Tube nucleus (*n*)

When transferred to the stigma of a plant of the same species, the pollen grain germinates to produce a pollen tube. The tube nucleus moves down the tube first, followed by the generative nucleus which soon divides mitotically to give two male nuclei.

Exine

Intine

Pollen tube

Male nuclei (*n*)

Tube nucleus (*n*)

Fig 12.5 Structure and development of the pollen grain

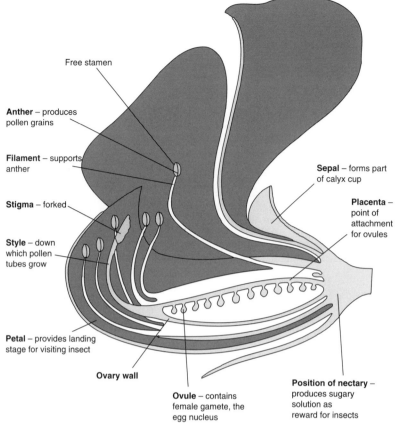

Free stamen

Anther – produces
pollen grains

Filament – supports
anther

Stigma – forked

Style – down
which pollen
tubes grow

Petal – provides landing
stage for visiting insect

Ovary wall

Ovule – contains
female gamete, the
egg nucleus

Sepal – forms part
of calyx cup

Placenta –
point of
attachment
for ovules

Position of nectary –
produces sugary
solution as
reward for insects

Fig 12.6 Sweet pea (*Lathyrus odoratus*)

12 Reproduction, development and growth

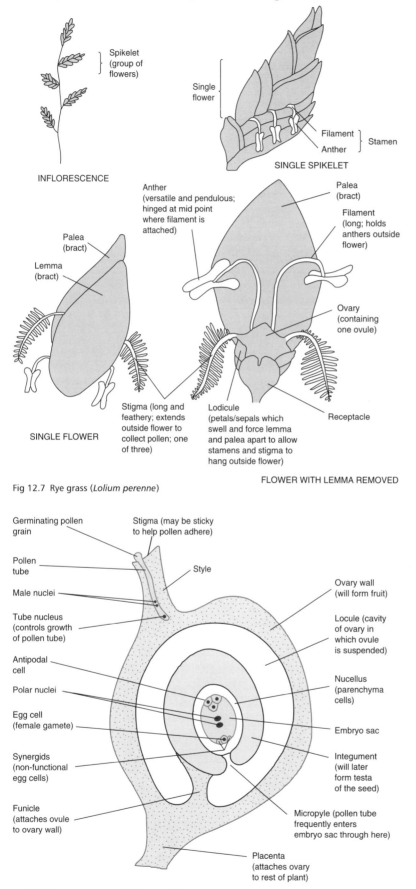

Spikelet (group of flowers)

INFLORESCENCE

Single flower

Filament
Anther
} Stamen

SINGLE SPIKELET

Anther (versatile and pendulous; hinged at mid point where filament is attached)

Palea (bract)

Filament (long; holds anthers outside flower)

Palea (bract)

Lemma (bract)

Ovary (containing one ovule)

Stigma (long and feathery; extends outside flower to collect pollen; one of three)

Lodicule (petals/sepals which swell and force lemma and palea apart to allow stamens and stigma to hang outside flower)

Receptacle

SINGLE FLOWER

FLOWER WITH LEMMA REMOVED

Fig 12.7 Rye grass (*Lolium perenne*)

Germinating pollen grain

Stigma (may be sticky to help pollen adhere)

Pollen tube

Style

Male nuclei

Ovary wall (will form fruit)

Tube nucleus (controls growth of pollen tube)

Locule (cavity of ovary in which ovule is suspended)

Antipodal cell

Polar nuclei

Nucellus (parenchyma cells)

Egg cell (female gamete)

Embryo sac

Synergids (non-functional egg cells)

Integument (will later form testa of the seed)

Funicle (attaches ovule to ovary wall)

Micropyle (pollen tube frequently enters embryo sac through here)

Placenta (attaches ovary to rest of plant)

Fig 12.9 Mature carpel at fertilization (LS)

12 Reproduction, development and growth

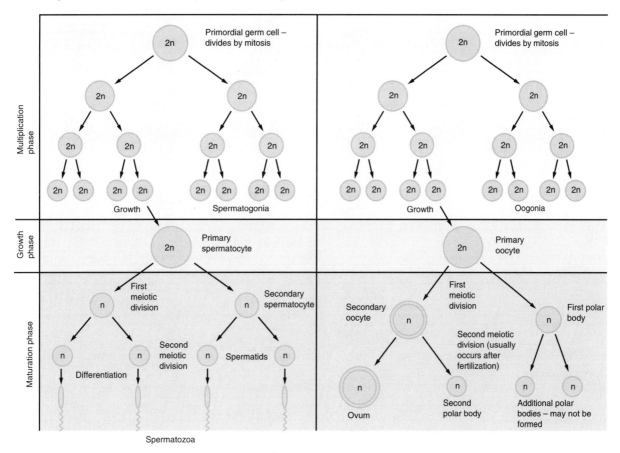

Fig 12.11 (a) Spermatogenesis – formation of sperm   (b) Oogenesis – formation of ova

12 Reproduction, development and growth

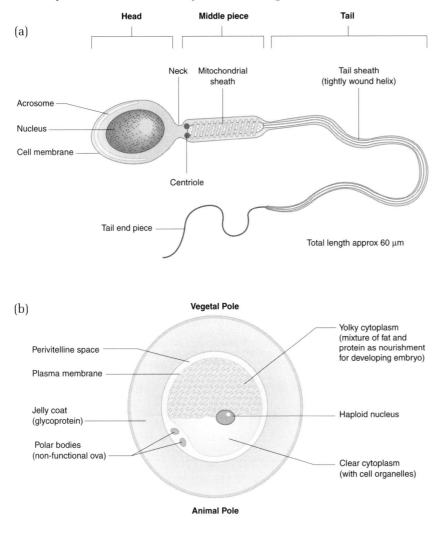

Fig 12.12 (a) Human spermatozoan based on electron micrograph (b) A generalized egg cell

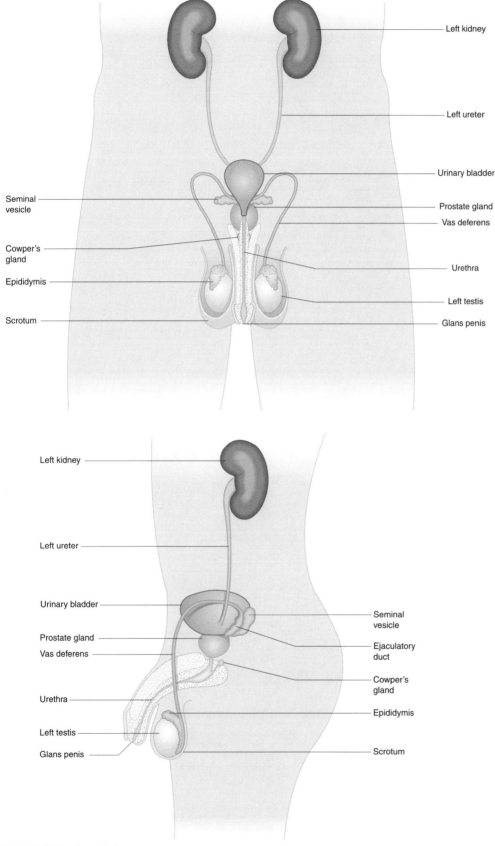

Fig 12.13 Male urinogenital system

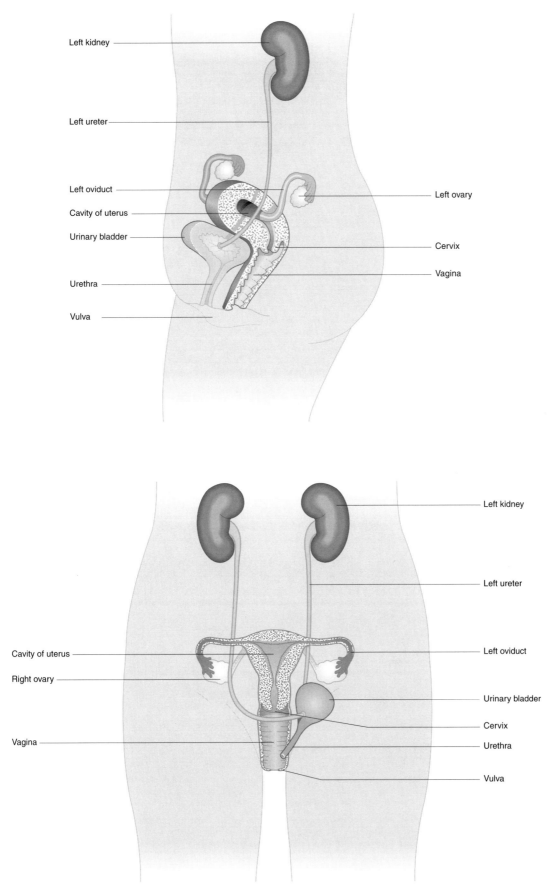

Fig 12.17 Female urinogenital system

12 Reproduction, development and growth

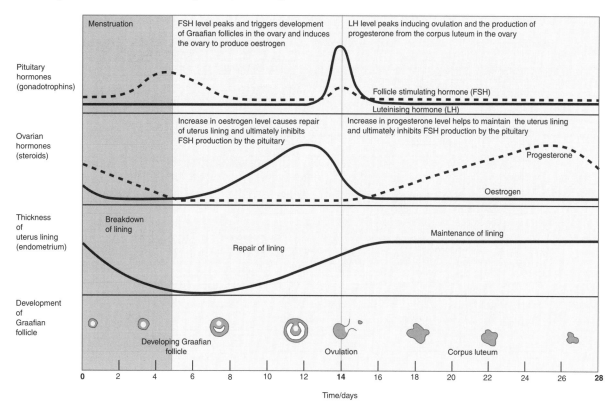

Fig 12.19 Diagram summarizing the events of the menstrual cycle

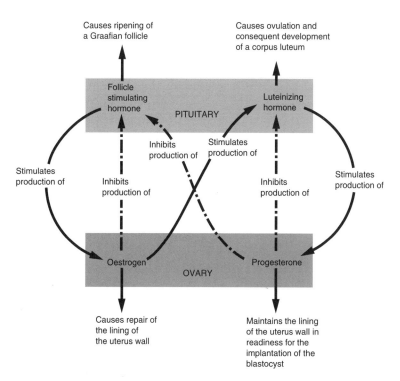

Fig 12.20 Hormone interaction in the menstrual cycle

12 Reproduction, development and growth

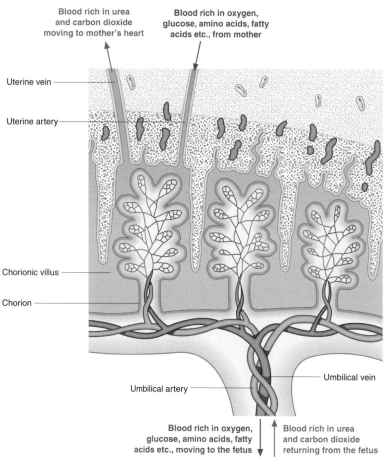

Blood rich in urea and carbon dioxide moving to mother's heart

Blood rich in oxygen, glucose, amino acids, fatty acids etc., from mother

Uterine vein

Uterine artery

Chorionic villus

Chorion

Umbilical vein

Umbilical artery

Blood rich in oxygen, glucose, amino acids, fatty acids etc., moving to the fetus

Blood rich in urea and carbon dioxide returning from the fetus

Fig 12.22 The mammalian placenta

The chorionic villi present a large surface area for the exchange of materials by diffusion across the chorionic membrane. In some mammals the maternal and fetal bloods flow in opposite directions., This counter-current flow leads to more efficient exchange as described in Section 20.2.4

14 Autotrophic nutrition (photosynthesis)

(a)

A

B

(b)

Section A/B

(c)

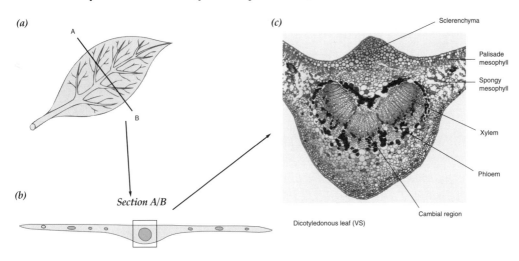

Sclerenchyma

Palisade mesophyll

Spongy mesophyll

Xylem

Phloem

Cambial region

Dicotyledonous leaf (VS)

(d) Dicotyledonous leaf (VS) (×40 approx.)

(e) Palisade cell

Cellulose cell wall

Chloroplasts

Cytoplasm

Vacuole

Nucleus

Tonoplast

Cell membrane

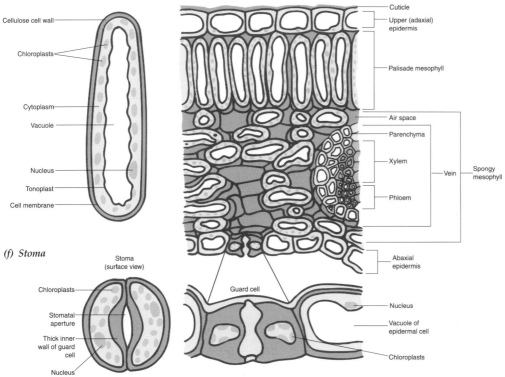

Cuticle

Upper (adaxial) epidermis

Palisade mesophyll

Air space

Parenchyma

Xylem

Phloem

Vein

Spongy mesophyll

Abaxial epidermis

Nucleus

Vacuole of epidermal cell

Chloroplasts

(f) Stoma

Stoma (surface view)

Chloroplasts

Stomatal aperture

Thick inner wall of guard cell

Nucleus

Guard cell

Fig 14.1 The structure of the leaf

14 Autotrophic nutrition (photosynthesis)

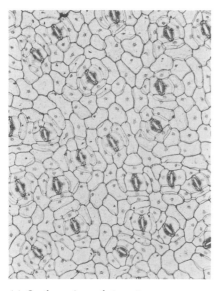

(g) Surface view of stomata

(h)

Stoma (VS)

Guard cell

Stoma

Sub-stomatal
air space

Fig 14.1 The structure of the leaf (continued)

84

14 Autotrophic nutrition (photosynthesis)

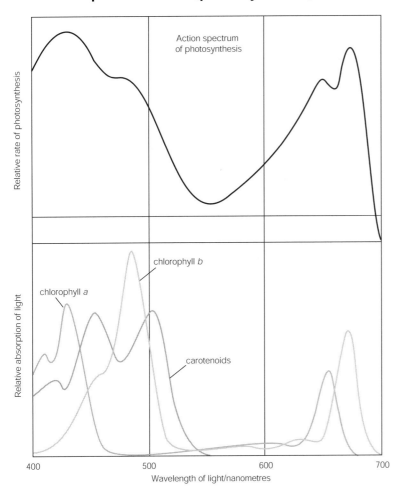

Fig 14.6 Action spectrum for photosynthesis and absorption spectra for common plant pigments

14 Autotrophic nutrition (photosynthesis)

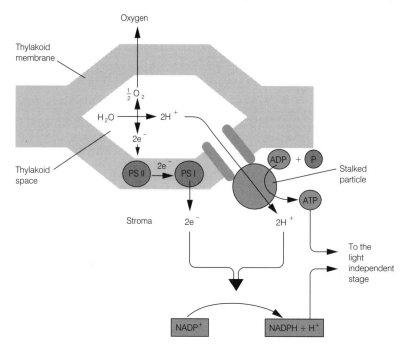

Fig 14.7 Summary of the events of the light dependent stage and their locations within the chloroplast

1. Light energy is trapped in photosystem II and boosts electrons to a higher energy level.
2. The electrons are received by an electron acceptor.
3. The electrons are passed from the electron acceptor along a series of electron carriers to photosystem I. The energy lost by the electrons is captured by converting ADP to ATP. Light energy has thereby been converted to chemical energy.
4. Light energy absorbed by photosystem I boosts the electrons to an even higher energy level.
5. The electrons are received by another electron acceptor.
6. The electrons which have been removed from the chlorophyll are replaced by pulling in other electrons from a water molecule.
7. The loss of electrons from the water molecule causes it to dissociate into protons and oxygen gas.
8. The protons from the water molecule combine with the electrons from the second electron acceptor and these reduce **nicotinamide adenine dinucleotide phosphate**.
9. Some electrons from the second acceptor may pass back to the chlorophyll molecule by the electron carrier system, yielding ATP as they do so. This process is called **cyclic photophosphorylation**.

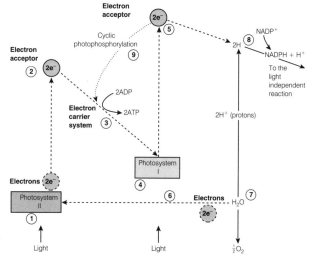

Fig 14.8 Summary of the light dependent stage of photosynthesis

14 Autotrophic nutrition (photosynthesis)

1. *Carbon dioxide diffuses into the leaf through the stomata and dissolves in the moisture on the walls of the palisade cells. It diffuses through the cell membrane, cytoplasm and chloroplast membrane into the stroma of the chloroplast.*
2. *The carbon dioxide combines with a 5-carbon compound called **ribulose bisphosphate** to form an unstable 6-carbon intermediate.*
3. *The 6-carbon intermediate breaks down into two molecules of the 3-carbon **glycerate 3-phosphate (GP)**.*
4. *Some of the ATP produced during the light dependent stage is used to help convert GP into **triose phosphate** (glyceraldehyde 3-phosphate – GALP).*
5. *The reduced nicotinamide adenine dinucleotide phosphate (NADPH+H⁺) from the light dependent reaction is necessary for the reduction of the GP to triose phosphate. NADP⁺ is regenerated and this returns to the light dependent stage to accept more hydrogen.*
6. *Pairs of triose phosphate molecules are combined to produce an intermediate hexose sugar.*
7. *The hexose sugar is polymerized to form starch which is stored by the plant.*
8. *Not all triose phosphate is combined to form starch. A portion of it is used to regenerate the original carbon dioxide acceptor, ribulose bisphosphate. Five molecules of the 3-carbon triose phosphate can regenerate three molecules of the 5-carbon ribulose bisphosphate. More of the ATP from the light dependent reaction is needed to provide the energy for this conversion.*

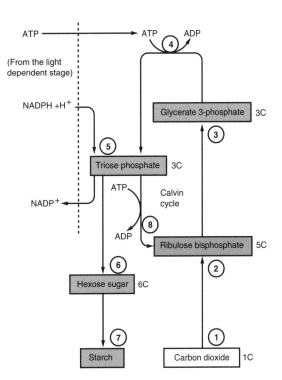

Fig 14.9 Summary of the light independent stage of photosynthesis

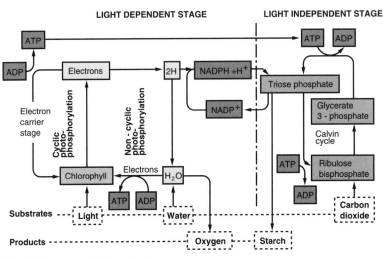

Fig 14.10 Summary of photosynthesis

15 Heterotrophic nutrition

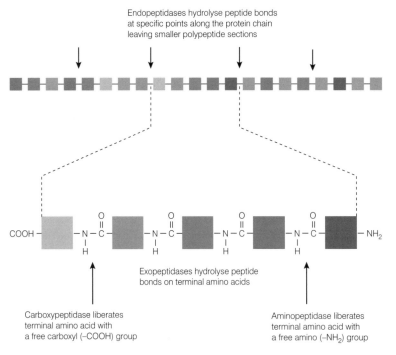

Endopeptidases hydrolyse peptide bonds
at specific points along the protein chain
leaving smaller polypeptide sections

Exopeptidases hydrolyse peptide
bonds on terminal amino acids

Carboxypeptidase liberates
terminal amino acid with
a free carboxyl (–COOH) group

Aminopeptidase liberates
terminal amino acid with
a free amino (–NH₂) group

Fig 15.1 Action of endo- and exopeptidases

15 Heterotrophic nutrition

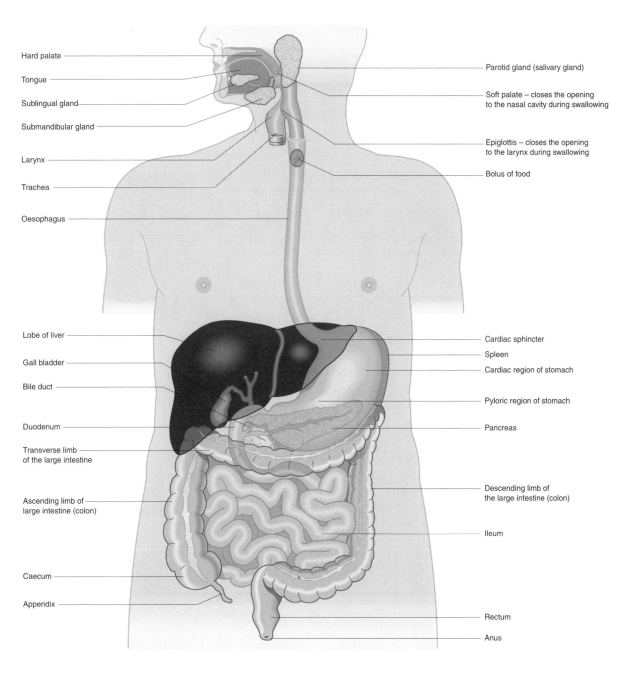

Hard palate

Tongue

Sublingual gland

Submandibular gland

Larynx

Trachea

Oesophagus

Lobe of liver

Gall bladder

Bile duct

Duodenum

Transverse limb
of the large intestine

Ascending limb of
large intestine (colon)

Caecum

Appendix

Parotid gland (salivary gland)

Soft palate – closes the opening
to the nasal cavity during swallowing

Epiglottis – closes the opening
to the larynx during swallowing

Bolus of food

Cardiac sphincter

Spleen

Cardiac region of stomach

Pyloric region of stomach

Pancreas

Descending limb of
the large intestine (colon)

Ileum

Rectum

Anus

Fig 15.2 Human digestive system

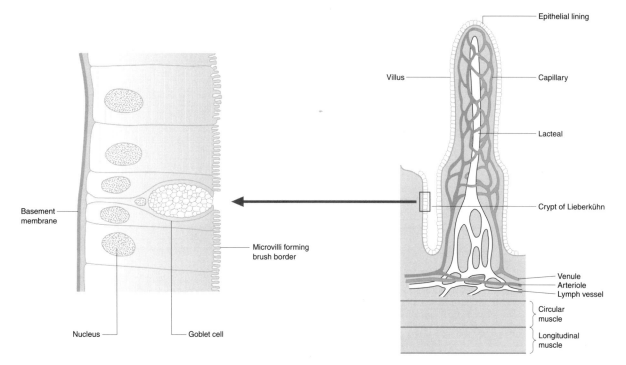

Fig 15.4 Intestinal wall showing villi (LS)

15 Heterotrophic nutrition

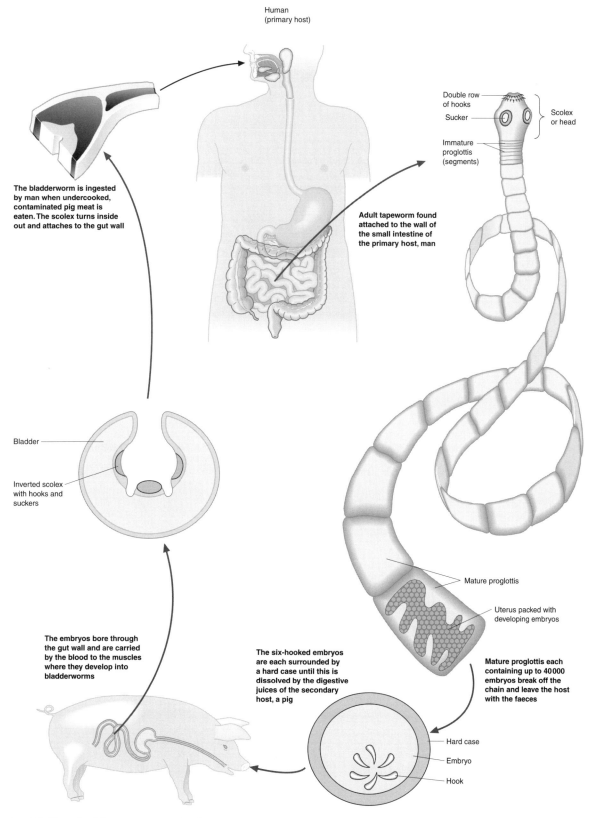

Human
(primary host)

Double row
of hooks

Sucker

Scolex
or head

Immature
proglottis
(segments)

The bladderworm is ingested
by man when undercooked,
contaminated pig meat is
eaten. The scolex turns inside
out and attaches to the gut wall

Adult tapeworm found
attached to the wall of
the small intestine of
the primary host, man

Bladder

Inverted scolex
with hooks and
suckers

Mature proglottis

Uterus packed with
developing embryos

The embryos bore through
the gut wall and are carried
by the blood to the muscles
where they develop into
bladderworms

The six-hooked embryos
are each surrounded by
a hard case until this is
dissolved by the digestive
juices of the secondary
host, a pig

Mature proglottis each
containing up to 40000
embryos break off the
chain and leave the host
with the faeces

Hard case

Embryo

Hook

Fig 15.5 Life cycle of the tapeworm *Taenia solium*

16 Cellular respiration

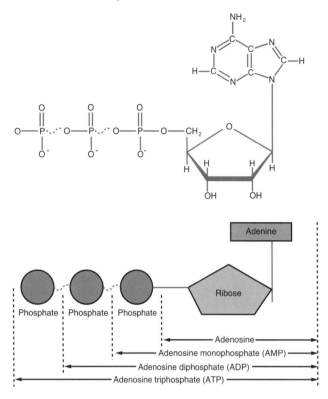

Fig 16.2 Structure of adenosine triphosphate

16 Cellular respiration

Stages of glycolysis

Glucose (six-carbon sugar) ATP → ADP	**Stage 1** The glucose molecule is phosphorylated to make it more reactive. The phosphate molecule comes from the conversion of ATP to ADP.
Glucose phosphate (6 C)	**Stage 2** The glucose molecule is reorganized into its isomer, fructose phosphate.
Fructose phosphate (6 C) ATP → ADP	**Stage 3** Further phosphorylation takes place, by the donation of another phosphate molecule from ATP to make the sugar yet more reactive.
Fructose bisphosphate (6 C)	**Stage 4** The six-carbon sugar is split into two three-carbon sugars.
Glyceraldehyde 3-phosphate (3 C) (2 molecules) Inorganic phosphate → $2 \times 2H$ Glycerate 1,3-bisphosphate (3 C) (2 molecules)	**Stage 5** More phosphorylation occurs but in this case the source of the phosphate is inorganic and not ATP. Two pairs of hydrogen atoms are removed.
2 ADP → 2 ATP Glycerate 3-phosphate (3 C) (2 molecules)	**Stage 6** A phosphate molecule is lost from both molecules of glycerate 1,3-bisphosphate, thus yielding two molecules of ATP from ADP.
2 ADP → 2 ATP → $2 H_2O$ Pyruvate (3 C) (2 molecules)	**Stage 7** A further pair of phosphates are removed forming two more ATPs. Each glycerate 3-phosphate molecule also has a water molecule removed.

16 Cellular respiration

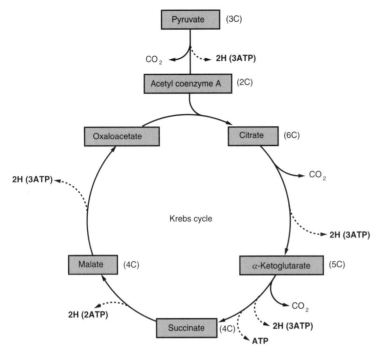

Fig 16.3 Summary of Krebs cycle

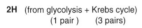

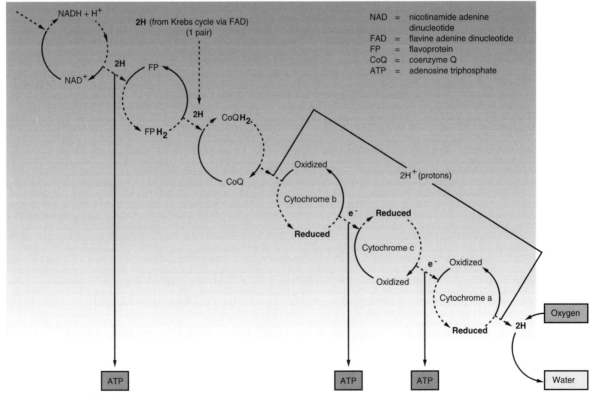

Fig 16.4 Summary of the electron transport system

16 Cellular respiration

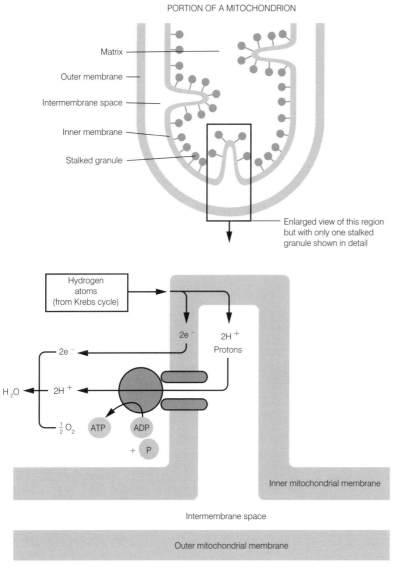

PORTION OF A MITOCHONDRION

Matrix

Outer membrane

Intermembrane space

Inner membrane

Stalked granule

Enlarged view of this region but with only one stalked granule shown in detail

Hydrogen atoms (from Krebs cycle)

$2e^-$

$2H^+$ Protons

$2e^-$

H_2O

$2H^+$

$\frac{1}{2}O_2$

ATP

ADP

+ P

Inner mitochondrial membrane

Intermembrane space

Outer mitochondrial membrane

Fig 16.5 The synthesis of ATP according to the chemi-osmotic theory of Mitchell

16 Cellular respiration

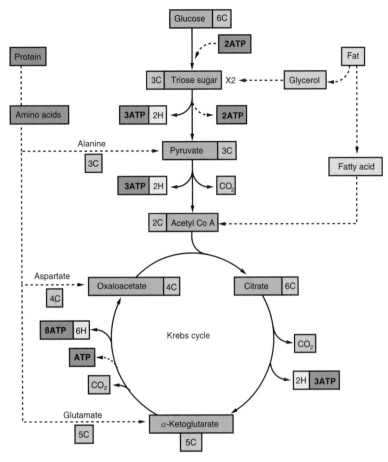

Fig 16.6 Summary of respiratory pathways

17 Energy and the ecosystem

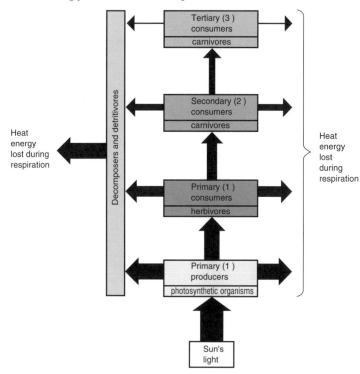

Fig 17.1 Energy flow through different trophic levels of a food chain

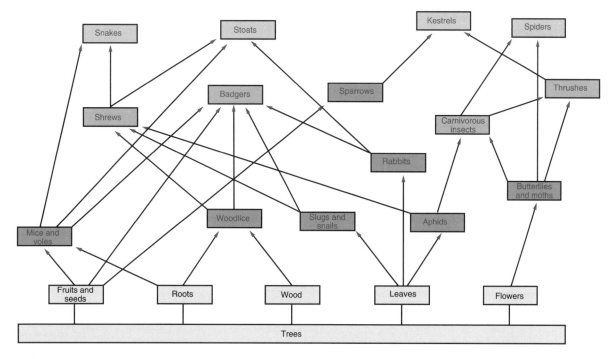

Fig 17.2 Simplified food web based on a woodland habitat

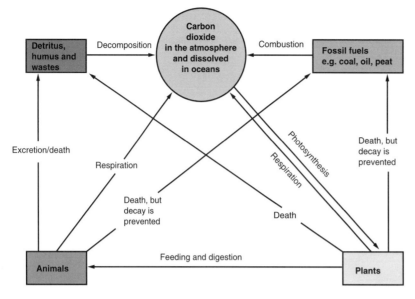

Fig 17.4 The carbon cycle

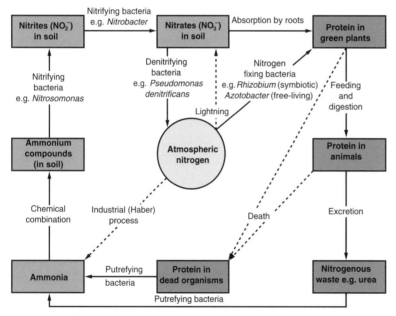

Fig 17.5 The nitrogen cycle

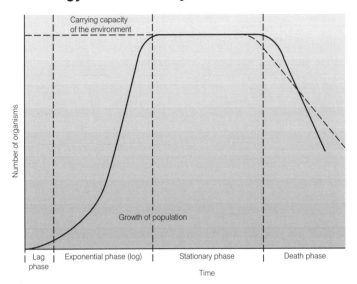

Fig 17.13 Growth of a population

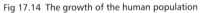
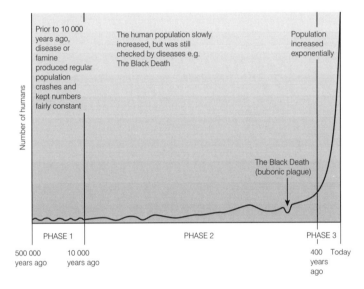

Fig 17.14 The growth of the human population

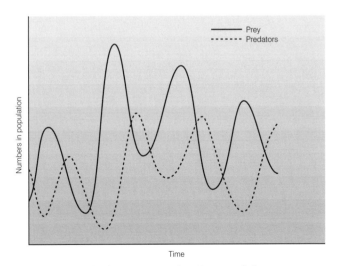

Fig 17.15 Relationship between prey and predator populations

17 Energy and the ecosystem

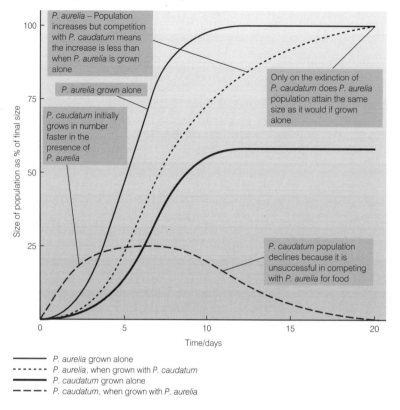

P. aurelia – Population increases but competition with *P. caudatum* means the increase is less than when *P. aurelia* is grown alone

P. aurelia grown alone

P. caudatum initially grows in number faster in the presence of *P. aurelia*

Only on the extinction of *P. caudatum* does *P. aurelia* population attain the same size as it would if grown alone

P. caudatum population declines because it is unsuccessful in competing with *P. aurelia* for food

——————— *P. aurelia* grown alone
- - - - - - - *P. aurelia*, when grown with *P. caudatum*
━━━━━━━ *P. caudatum* grown alone
— — — — *P. caudatum*, when grown with *P. aurelia*

Fig 17.16 Population growth of two species of *Paramecium* grown separately and together

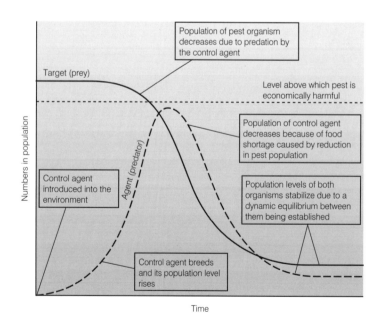

Population of pest organism decreases due to predation by the control agent

Target (prey)

Level above which pest is economically harmful

Population of control agent decreases because of food shortage caused by reduction in pest population

Control agent introduced into the environment

Agent (predator)

Population levels of both organisms stabilize due to a dynamic equilibrium between them being established

Control agent breeds and its population level rises

Fig 17.17 General relationships between pest and control agent populations in biological control

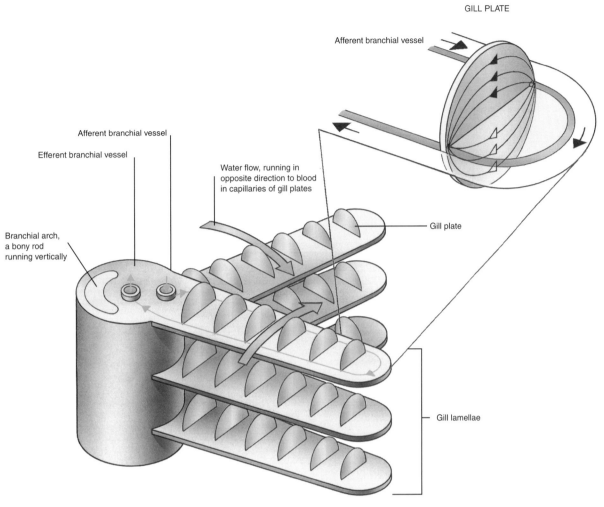

GILL PLATE

Afferent branchial vessel

Afferent branchial vessel

Efferent branchial vessel

Water flow, running in opposite direction to blood in capillaries of gill plates

Branchial arch, a bony rod running vertically

Gill plate

Gill lamellae

Fig 20.3 Water flow over gill lamellae in a bony fish

20 Gaseous exchange

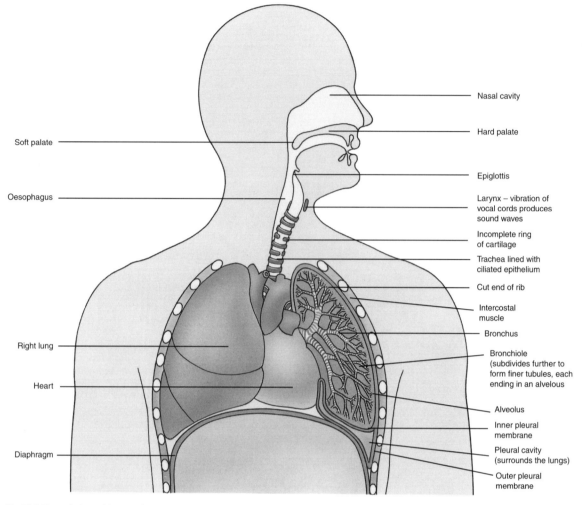

Nasal cavity

Hard palate

Soft palate

Epiglottis

Oesophagus

Larynx – vibration of
vocal cords produces
sound waves

Incomplete ring
of cartilage

Trachea lined with
ciliated epithelium

Cut end of rib

Intercostal
muscle

Right lung

Bronchus

Heart

Bronchiole
(subdivides further to
form finer tubules, each
ending in an alvelous

Alveolus

Inner pleural
membrane

Pleural cavity
(surrounds the lungs)

Diaphragm

Outer pleural
membrane

Fig 20.6 Ventral view of human thorax

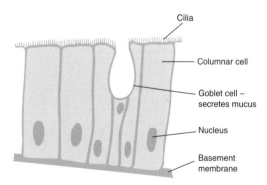

Cilia

Columnar cell

Goblet cell –
secretes mucus

Nucleus

Basement
membrane

Fig 20.7 Ciliated epithelium (LS)

20 Gaseous exchange

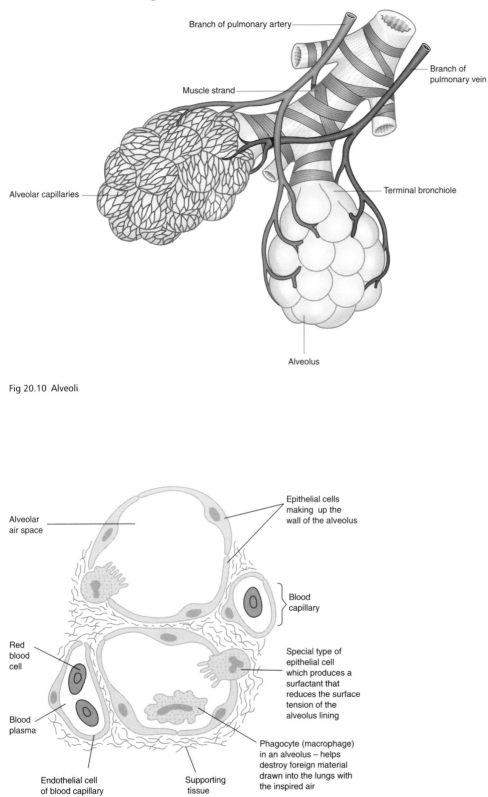

Fig 20.10 Alveoli

Fig 20.12 Arrangement of cells and tissues in mammalian lung alveoli

20 Gaseous exchange

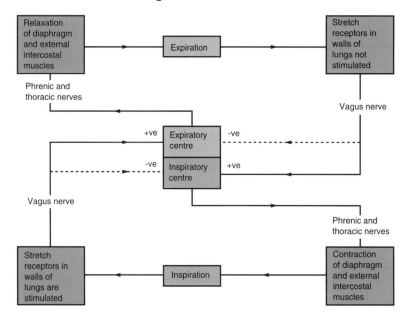

Fig 20.13 Control of ventilation

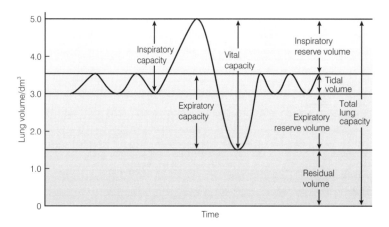

Fig 20.15 Graph to illustrate lung capacities

21 Blood and circulation (transport in animals)

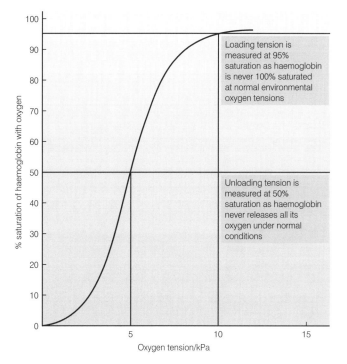

Loading tension is measured at 95% saturation as haemoglobin is never 100% saturated at normal environmental oxygen tensions

Unloading tension is measured at 50% saturation as haemoglobin never releases all its oxygen under normal conditions

Fig 21.3 Oxygen dissociation curve for adult haemoglobin

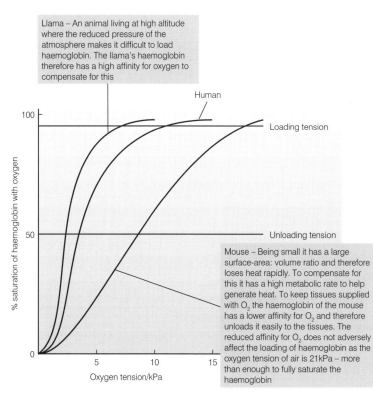

Llama – An animal living at high altitude where the reduced pressure of the atmosphere makes it difficult to load haemoglobin. The llama's haemoglobin therefore has a high affinity for oxygen to compensate for this

Human

Loading tension

Unloading tension

Mouse – Being small it has a large surface-area: volume ratio and therefore loses heat rapidly. To compensate for this it has a high metabolic rate to help generate heat. To keep tissues supplied with O_2 the haemoglobin of the mouse has a lower affinity for O_2 and therefore unloads it easily to the tissues. The reduced affinity for O_2 does not adversely affect the loading of haemoglobin as the oxygen tension of air is 21kPa – more than enough to fully saturate the haemoglobin

Fig 21.4 Oxygen dissociation curves for the haemoglobin of three mammals

21 Blood and circulation (transport in animals)

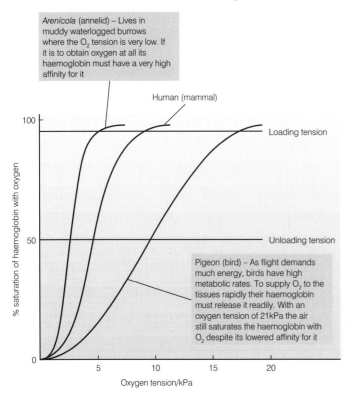

Arenicola (annelid) – Lives in muddy waterlogged burrows where the O₂ tension is very low. If it is to obtain oxygen at all its haemoglobin must have a very high affinity for it

Human (mammal)

Loading tension

Unloading tension

Pigeon (bird) – As flight demands much energy, birds have high metabolic rates. To supply O₂ to the tissues rapidly their haemoglobin must release it readily. With an oxygen tension of 21kPa the air still saturates the haemoglobin with O₂ despite its lowered affinity for it

% saturation of haemoglobin with oxygen

Oxygen tension/kPa

Fig 21.5 Oxygen dissociation curves for the haemoglobin of three animals from different groups

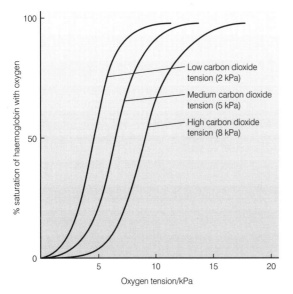

% saturation of haemoglobin with oxygen

Low carbon dioxide tension (2 kPa)

Medium carbon dioxide tension (5 kPa)

High carbon dioxide tension (8 kPa)

Oxygen tension/kPa

Fig 21.6 Oxygen dissociation curve of human haemoglobin, illustrating the Bohr effect

21 Blood and circulation (transport in animals)

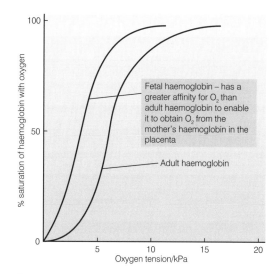

Fig 21.7 Comparison of the oxygen dissociation curves of adult and fetal haemoglobin

The following labels appear on the graph:

% saturation of haemoglobin with oxygen (y-axis)

Oxygen tension/kPa (x-axis)

Fetal haemoglobin – has a greater affinity for O_2 than adult haemoglobin to enable it to obtain O_2 from the mother's haemoglobin in the placenta

Adult haemoglobin

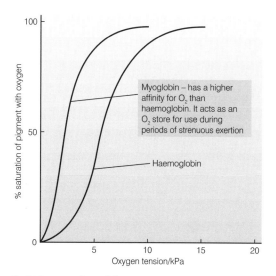

Fig 21.8 Comparison of the oxygen dissociation curves of human haemoglobin and myoglobin

The following labels appear on the graph:

% saturation of pigment with oxygen (y-axis)

Oxygen tension/kPa (x-axis)

Myoglobin – has a higher affinity for O_2 than haemoglobin. It acts as an O_2 store for use during periods of strenuous exertion

Haemoglobin

21 Blood and circulation (transport in animals)

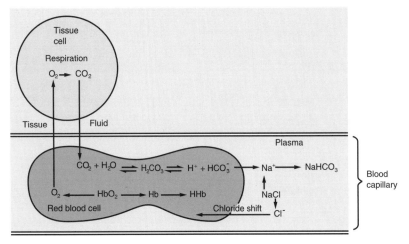

Fig 21.9 The chloride shift

The variable region differs with each antibody. It has a shape which exactly fits an antigen. Each antibody therefore can bind to two antigens.

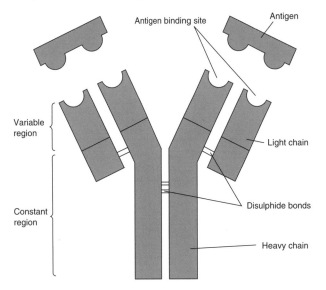

Fig 21.12 Structure of an antibody

21 Blood and circulation (transport in animals)

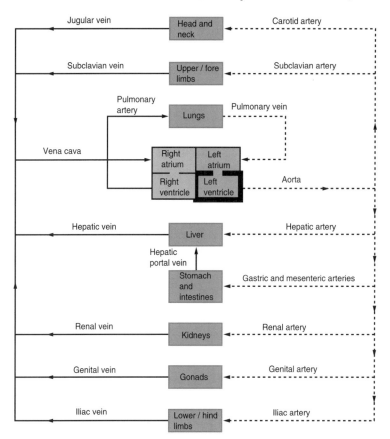

Fig 21.15 General plan of the mammalian circulatory system

21 Blood and circulation (transport in animals)

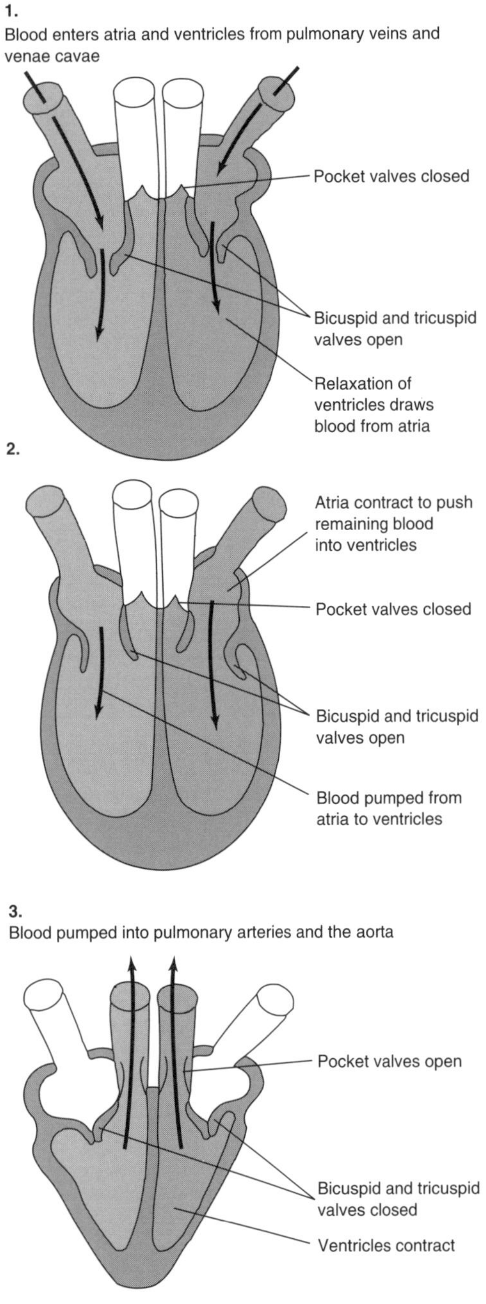

1.
Blood enters atria and ventricles from pulmonary veins and venae cavae

Pocket valves closed

Bicuspid and tricuspid valves open

Relaxation of ventricles draws blood from atria

2.

Atria contract to push remaining blood into ventricles

Pocket valves closed

Bicuspid and tricuspid valves open

Blood pumped from atria to ventricles

3.
Blood pumped into pulmonary arteries and the aorta

Pocket valves open

Bicuspid and tricuspid valves closed

Ventricles contract

Fig 21.17 The cardiac cycle

1. *Diastole*
   Atria are relaxed and fill with blood. Ventricles are also relaxed.
2. *Atrial systole*
   Atria contract pushing blood into the ventricles.
   Ventricles remain relaxed.
3. *Ventricular systole*
   Atria relax. Ventricles contract pushing blood away from heart through pulmonary arteries and the aorta.

21 Blood and circulation (transport in animals)

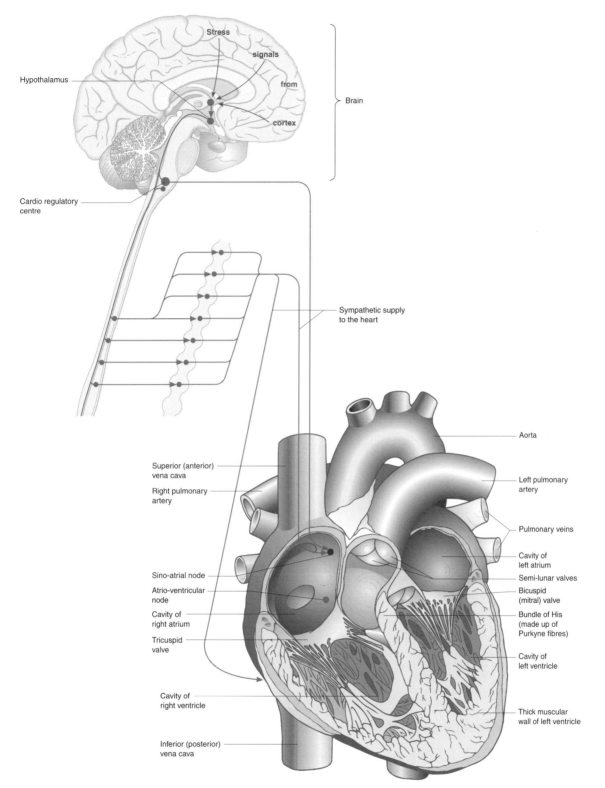

Fig 21.19 VS through mammalian heart to show position of sino-atrial node, atrio-ventricular node, bundle of His and nervous connections to the brain

21 Blood and circulation (transport in animals)

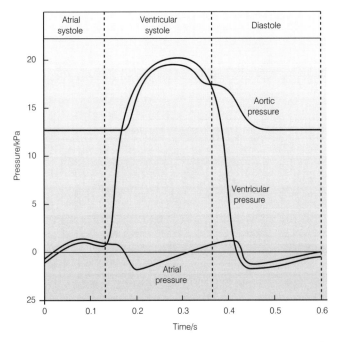

Fig 21.20 Pressure changes in the atria, ventricles and aorta during one cardiac cycle

21 Blood and circulation (transport in animals)

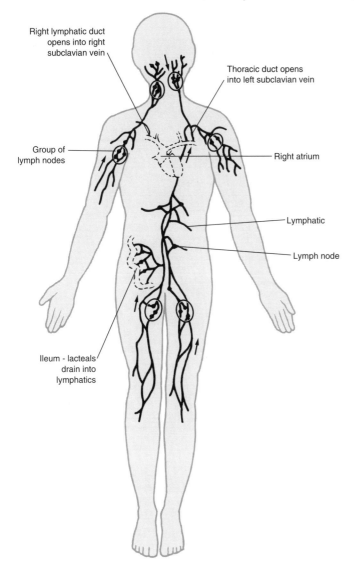

Right lymphatic duct opens into right subclavian vein

Thoracic duct opens into left subclavian vein

Group of lymph nodes

Right atrium

Lymphatic

Lymph node

Ileum - lacteals drain into lymphatics

Fig 21.22 The human lymphatic system

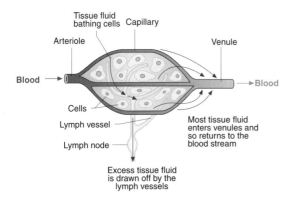

Tissue fluid bathing cells

Capillary

Arteriole

Venule

Blood

Blood

Cells

Lymph vessel

Most tissue fluid enters venules and so returns to the blood stream

Lymph node

Excess tissue fluid is drawn off by the lymph vessels

Fig 21.23 Formation and destination of tissue fluid

23 Osmoregulation and excretion

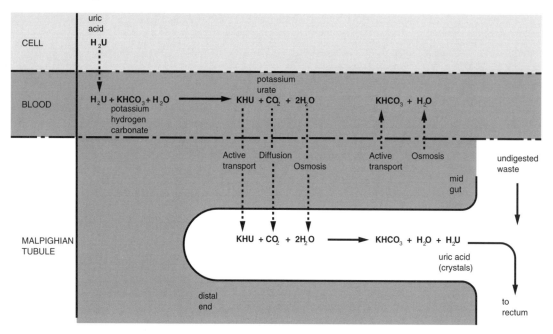

Fig 23.2 Functioning of the Malpighian tubule

23 Osmoregulation and excretion

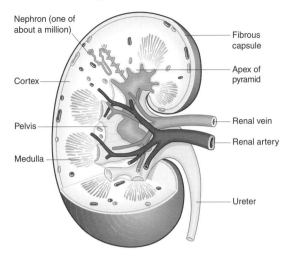

Nephron (one of about a million)

Cortex

Pelvis

Medulla

Fibrous capsule

Apex of pyramid

Renal vein

Renal artery

Ureter

Fig 23.4 Mammalian kidney to show position of a nephron (LS)

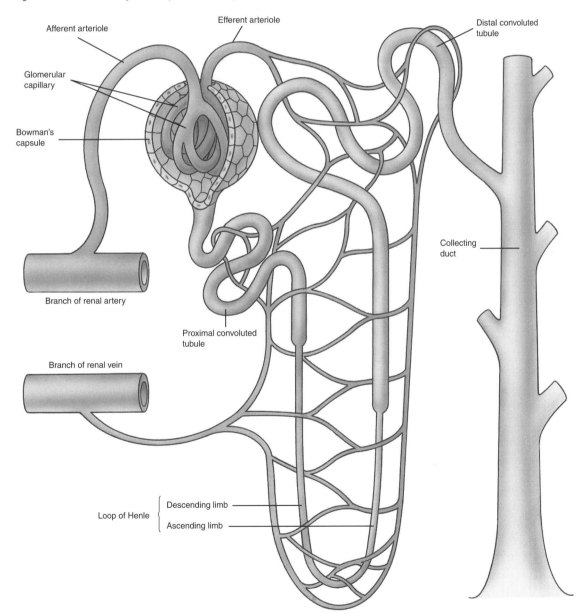

Afferent arteriole

Efferent arteriole

Distal convoluted tubule

Glomerular capillary

Bowman's capsule

Branch of renal artery

Branch of renal vein

Proximal convoluted tubule

Collecting duct

Loop of Henle { Descending limb
Ascending limb

Fig 23.5 Regions of the nephron

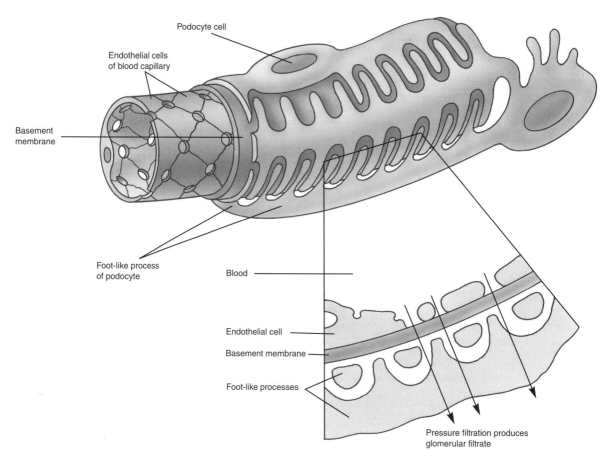

Podocyte cell

Endothelial cells
of blood capillary

Basement
membrane

Foot-like process
of podocyte

Blood

Endothelial cell

Basement membrane

Foot-like processes

Pressure filtration produces
glomerular filtrate

Fig 23.6 Podocyte

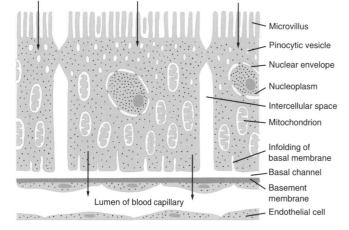

Lumen of proximal convoluted tubule

Reabsorption of material by diffusion, active transport and pinocytosis

Microvillus

Pinocytic vesicle

Nuclear envelope

Nucleoplasm

Intercellular space

Mitochondrion

Infolding of
basal membrane

Basal channel

Basement
membrane

Endothelial cell

Lumen of blood capillary

Fig 23.7 Detail of cells from the wall of the proximal convoluted tubule

23 Osmoregulation and excretion

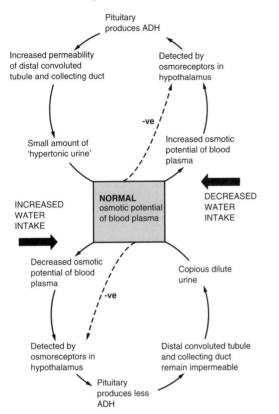

Fig 23.11 Regulation of ADH production

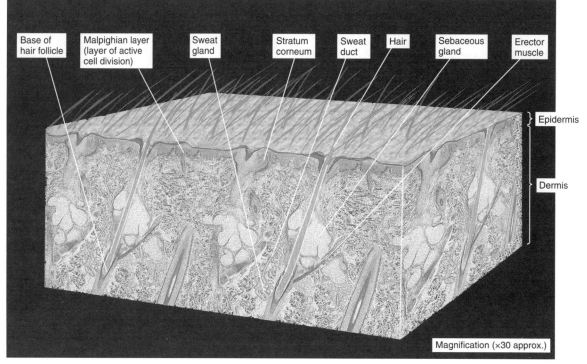

Fig 25.3 VS through human skin

25 Homeostasis

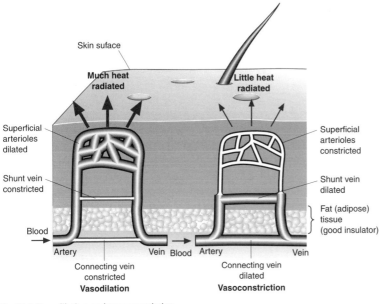

Fig 25.4 Vasodilation and vasoconstriction

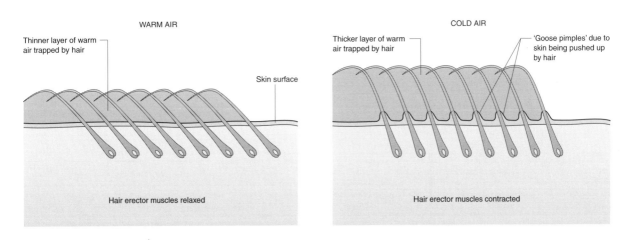

Fig 25.5 Lowering and raising of hair in controlling heat loss

25 Homeostasis

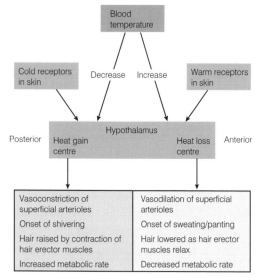

Fig 25.7 Summary of body temperature control by the hypothalamus

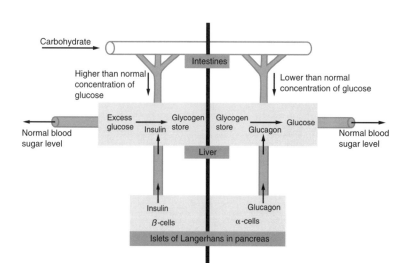

Fig 25.8 Summary of the control of blood sugar level

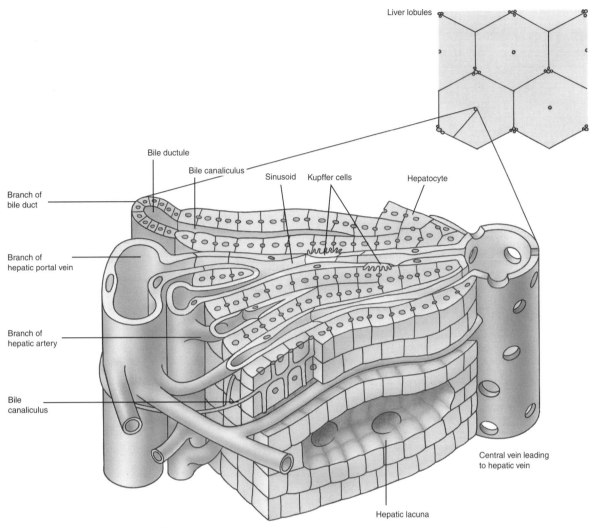

Liver lobules

Bile ductule

Bile canaliculus

Sinusoid

Kupffer cells

Hepatocyte

Branch of
bile duct

Branch of
hepatic portal vein

Branch of
hepatic artery

Bile
canaliculus

Central vein leading
to hepatic vein

Hepatic lacuna

Fig 25.10 Structure of the mammalian liver

26 The endocrine system

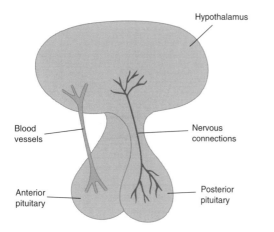

Fig 26.2 Structure of the pituitary gland and its relationship to the hypothalamus

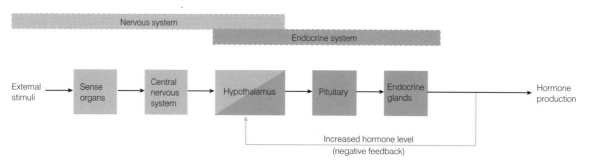

Fig 26.3 Role of the hypothalamus as the link between the nervous and the endocrine systems

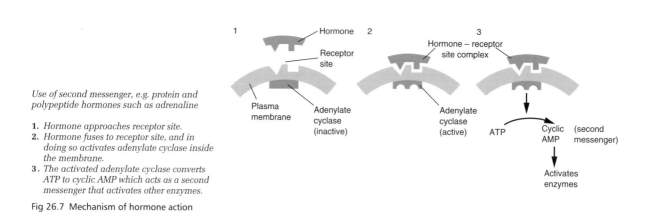

Use of second messenger, e.g. protein and polypeptide hormones such as adrenaline

1. *Hormone approaches receptor site.*
2. *Hormone fuses to receptor site, and in doing so activates adenylate cyclase inside the membrane.*
3. *The activated adenylate cyclase converts ATP to cyclic AMP which acts as a second messenger that activates other enzymes.*

Fig 26.7 Mechanism of hormone action

26 The endocrine system

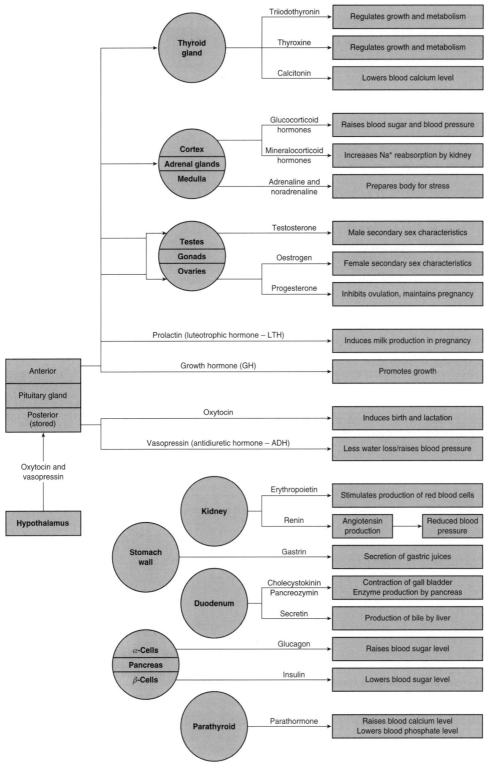

Fig 26.8 The endocrine system

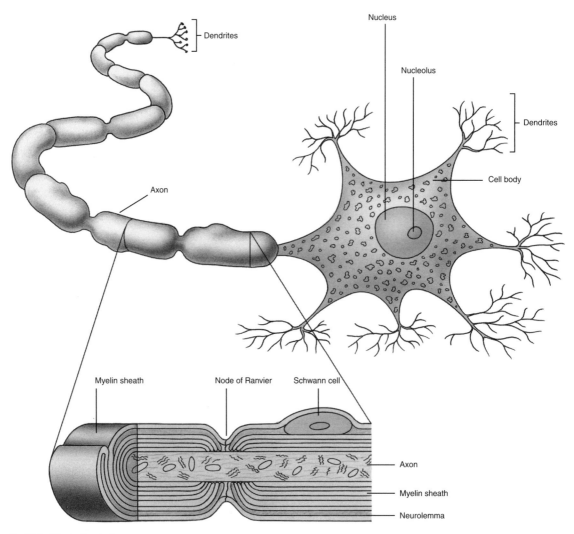

Fig 27.2 Effector (motor) neurone

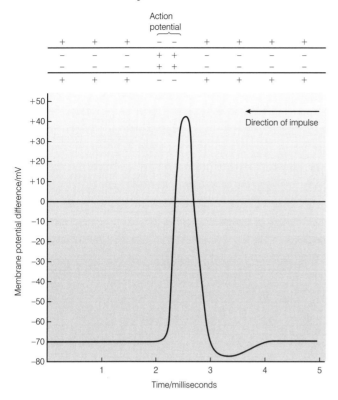

Fig 27.4 The action potential

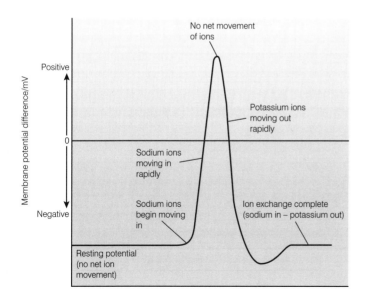

Fig 27.5 Ion movements during an action potential

27 The nervous system

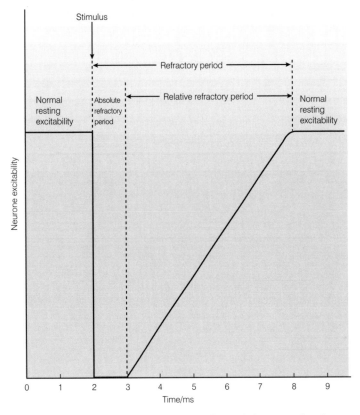

Fig 27.6 Graph illustrating neurone excitability before and after a nerve impulse

Where the stimulus is at the threshold value the excitability of the neurone must return to normal before a new action potential can be formed. In the time interval shown, this allows just two action potentials to pass, i.e. a low frequency of impulses. Where the stimulus exceeds the threshold value, a new action potential can be created before neurone excitability returns to normal. In the time interval shown, this allows six action potentials to pass, i.e. a high frequency of impulses.

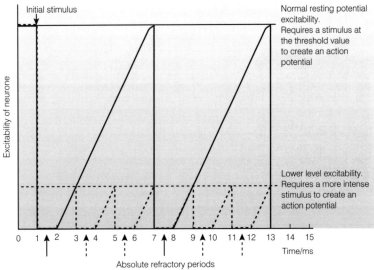

Fig 27.7 Determination of impulse frequency

27 The nervous system

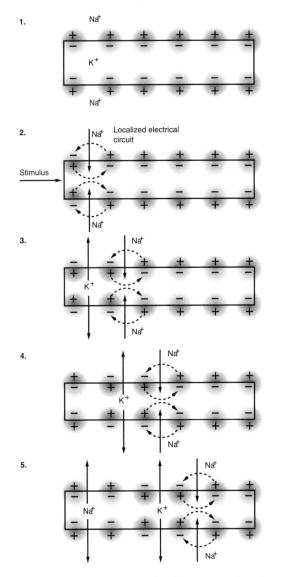

Fig 27.8 Transmission of an impulse along an unmyelinated neurone

1. At resting potential there is a high concentration of sodium ions outside and a high concentration of potassium ions inside the neurone.
2. When the neurone is stimulated sodium ions rush into the axon along a concentration gradient. This causes depolarization of the membrane.
3. Localized electrical circuits are established which cause further influx of sodium ions and so progression of the impulse. Behind the impulse, potassium ions begin to leave the axon along a concentration gradient.
4. As the impulse progresses, the outflux of potassium ions causes the neurone to become repolarized behind the impulse.
5. After the impulse has passed and the neurone is repolarized, sodium is once again actively expelled in order to increase the external concentration and so allow the passage of another impulse.

27 The nervous system

The insulating myelin causes ion exchange to occur at the nodes of Ranvier. The impulse therefore jumps from node to node.

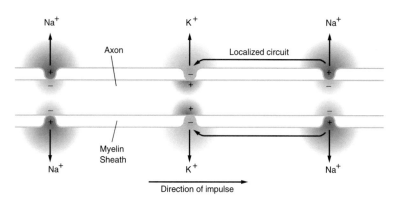

Fig 27.9 Transmission of an impulse along a myelinated neurone

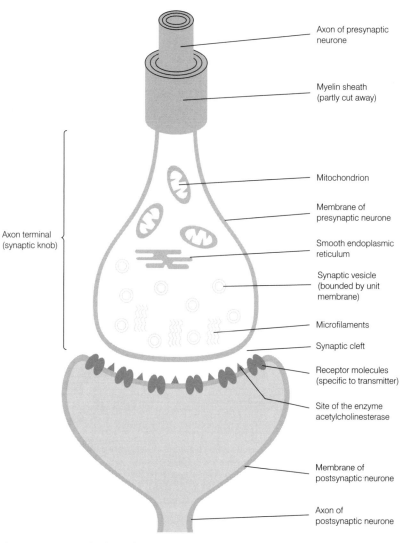

Fig 27.10 Structure of a chemical synapse

27 The nervous system

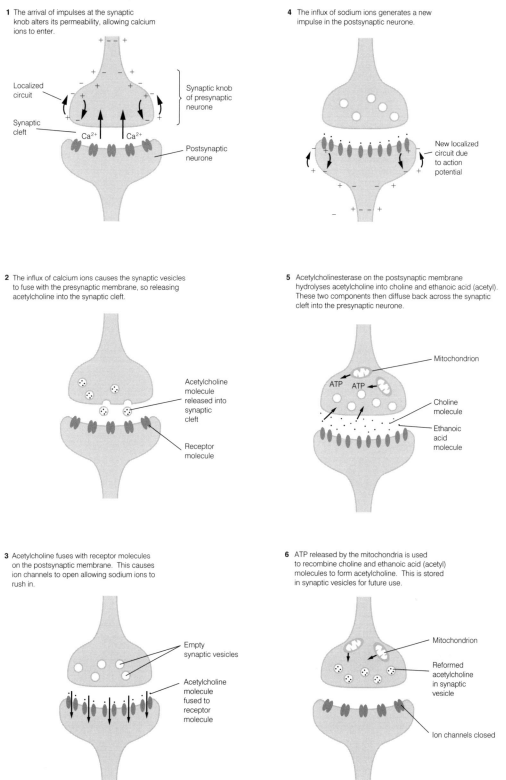

1 The arrival of impulses at the synaptic knob alters its permeability, allowing calcium ions to enter.

Localized circuit

Synaptic cleft

Synaptic knob of presynaptic neurone

Ca²⁺   Ca²⁺

Postsynaptic neurone

2 The influx of calcium ions causes the synaptic vesicles to fuse with the presynaptic membrane, so releasing acetylcholine into the synaptic cleft.

Acetylcholine molecule released into synaptic cleft

Receptor molecule

3 Acetylcholine fuses with receptor molecules on the postsynaptic membrane. This causes ion channels to open allowing sodium ions to rush in.

Empty synaptic vesicles

Acetylcholine molecule fused to receptor molecule

4 The influx of sodium ions generates a new impulse in the postsynaptic neurone.

New localized circuit due to action potential

5 Acetylcholinesterase on the postsynaptic membrane hydrolyses acetylcholine into choline and ethanoic acid (acetyl). These two components then diffuse back across the synaptic cleft into the presynaptic neurone.

Mitochondrion

ATP   ATP

Choline molecule

Ethanoic acid molecule

6 ATP released by the mitochondria is used to recombine choline and ethanoic acid (acetyl) molecules to form acetylcholine. This is stored in synaptic vesicles for future use.

Mitochondrion

Reformed acetylcholine in synaptic vesicle

Ion channels closed

Fig 27.11 Sequence of diagrams to illustrate synaptic transmission (only relevant detail is included in each drawing)

27 The nervous system

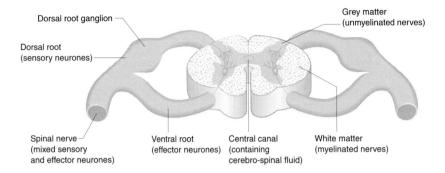

Fig 27.14 TS through the spinal cord

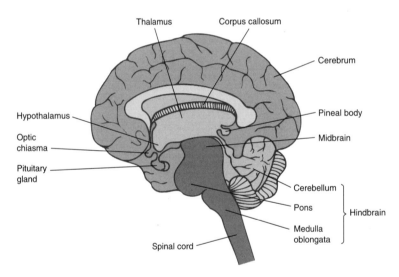

Fig 27.15 VS through the centre of the human brain

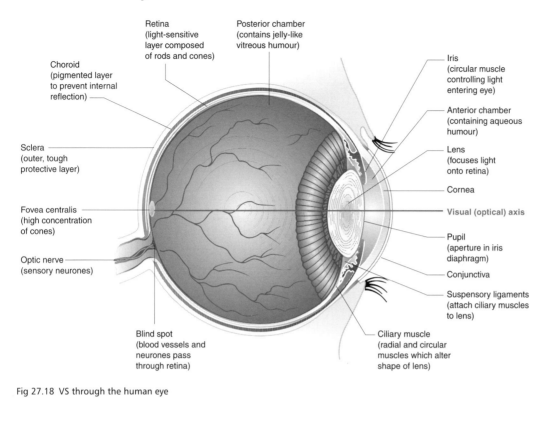

Retina
(light-sensitive layer composed of rods and cones)

Posterior chamber
(contains jelly-like vitreous humour)

Choroid
(pigmented layer to prevent internal reflection)

Iris
(circular muscle controlling light entering eye)

Anterior chamber
(containing aqueous humour)

Sclera
(outer, tough protective layer)

Lens
(focuses light onto retina)

Cornea

Visual (optical) axis

Fovea centralis
(high concentration of cones)

Pupil
(aperture in iris diaphragm)

Conjunctiva

Optic nerve
(sensory neurones)

Suspensory ligaments
(attach ciliary muscles to lens)

Blind spot
(blood vessels and neurones pass through retina)

Ciliary muscle
(radial and circular muscles which alter shape of lens)

Fig 27.18 VS through the human eye

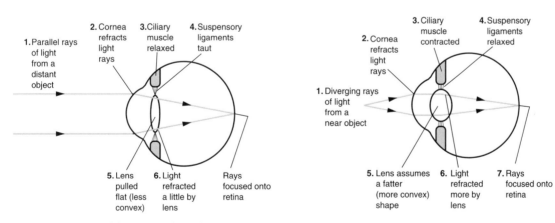

1. Parallel rays of light from a distant object
2. Cornea refracts light rays
3. Ciliary muscle relaxed
4. Suspensory ligaments taut
5. Lens pulled flat (less convex)
6. Light refracted a little by lens
Rays focused onto retina

1. Diverging rays of light from a near object
2. Cornea refracts light rays
3. Ciliary muscle contracted
4. Suspensory ligaments relaxed
5. Lens assumes a fatter (more convex) shape
6. Light refracted more by lens
7. Rays focused onto retina

Fig 27.20 (a) Condition of the eye when focused on a distant object   (b) Condition of the eye when focused on a near object

27 The nervous system

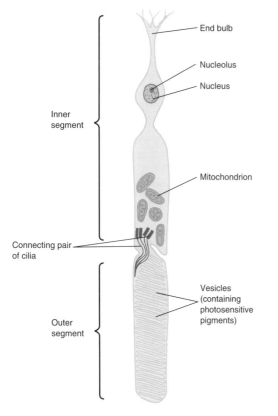

Fig 27.22 Structure of a single rod cell

Labels for Fig 27.22:
- End bulb
- Nucleolus
- Nucleus
- Inner segment
- Mitochondrion
- Connecting pair of cilia
- Vesicles (containing photosensitive pigments)
- Outer segment

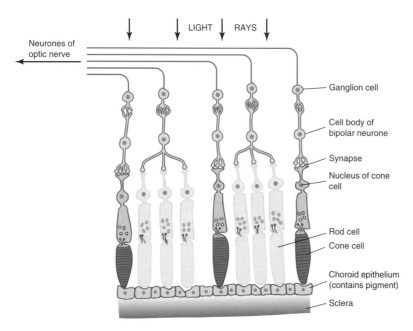

Fig 27.23 Microscopic structure of the retina

Labels for Fig 27.23:
- LIGHT RAYS
- Neurones of optic nerve
- Ganglion cell
- Cell body of bipolar neurone
- Synapse
- Nucleus of cone cell
- Rod cell
- Cone cell
- Choroid epithelium (contains pigment)
- Sclera

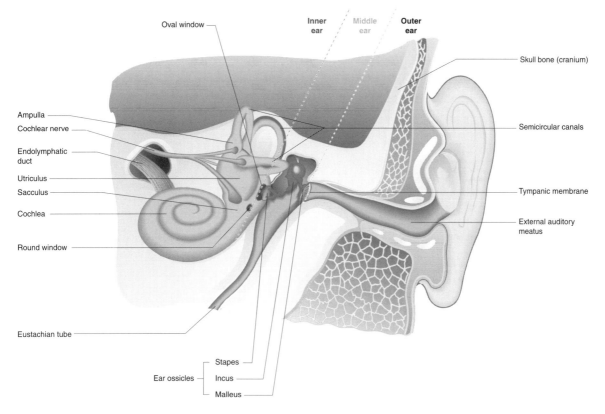

Fig 27.25 The human ear

27 The nervous system

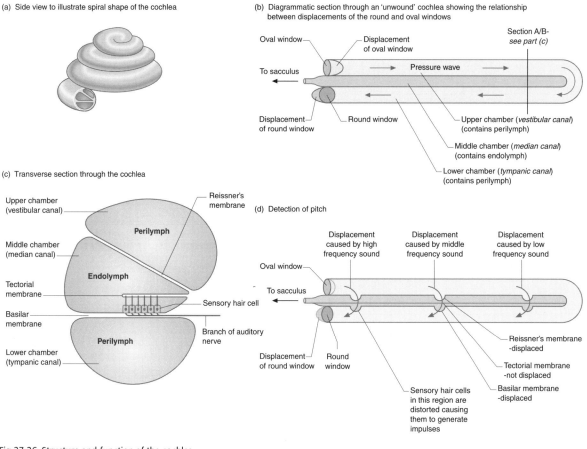

(a) Side view to illustrate spiral shape of the cochlea

(b) Diagrammatic section through an 'unwound' cochlea showing the relationship between displacements of the round and oval windows

Oval window
Displacement of oval window
Section A/B- see part (c)
To sacculus
Pressure wave
Displacement of round window
Round window
Upper chamber (*vestibular canal*) (contains perilymph)
Middle chamber (*median canal*) (contains endolymph)
Lower chamber (*tympanic canal*) (contains perilymph)

(c) Transverse section through the cochlea

Upper chamber (vestibular canal)
Reissner's membrane
Perilymph
Middle chamber (median canal)
Endolymph
Tectorial membrane
Sensory hair cell
Basilar membrane
Branch of auditory nerve
Perilymph
Lower chamber (tympanic canal)

(d) Detection of pitch

Displacement caused by high frequency sound
Displacement caused by middle frequency sound
Displacement caused by low frequency sound
Oval window
To sacculus
Reissner's membrane -displaced
Displacement of round window
Round window
Tectorial membrane -not displaced
Sensory hair cells in this region are distorted causing them to generate impulses
Basilar membrane -displaced

Fig 27.26 Structure and function of the cochlea

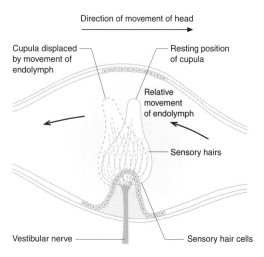

Direction of movement of head

Cupula displaced by movement of endolymph
Resting position of cupula
Relative movement of endolymph
Sensory hairs
Vestibular nerve
Sensory hair cells

Fig 27.27 Section through ampulla of a semi-circular canal

28 Muscular movement and support

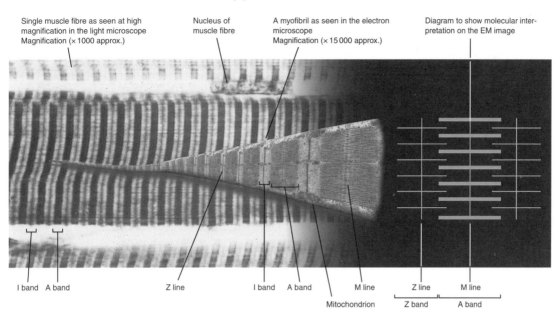

Single muscle fibre as seen at high magnification in the light microscope Magnification (× 1000 approx.)

Nucleus of muscle fibre

A myofibril as seen in the electron microscope Magnification (× 15 000 approx.)

Diagram to show molecular interpretation on the EM image

I band   A band

Z line

I band   A band

M line

Mitochondrion

Z line

M line

Z band

A band

Fig 28.1 Detailed structure of muscle

28 Muscular movement and support

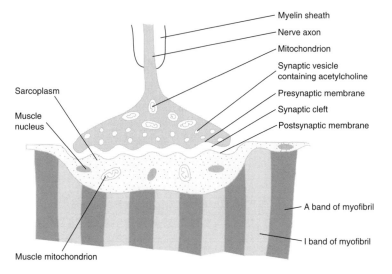

Fig 28.2 Neuromuscular junction – the end plate

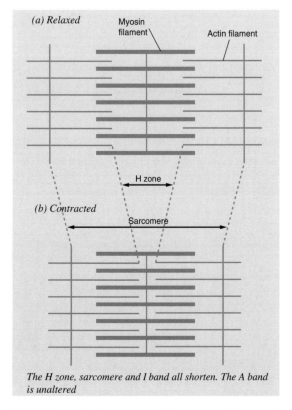

(a) Relaxed

Myosin filament

Actin filament

H zone

(b) Contracted

Sarcomere

The H zone, sarcomere and I band all shorten. The A band is unaltered

Fig 28.4 Changes in the appearance of a sarcomere during muscle contraction

28 Muscular movement and support

(a) The head of the myosin molecule is 'cocked' ready to attach to the actin filament

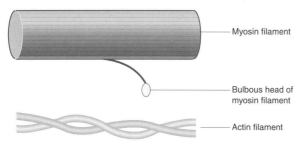

Myosin filament

Bulbous head of myosin filament

Actin filament

(b) Myosin head attaches to a monomer unit on the actin molecule

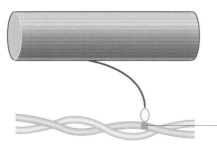

Monomer unit to which myosin head attaches

(c) The myosin changes position in order to attain a lower energy state. In doing so it slides the actin filament past the stationary myosin filament

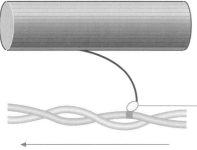

Myosin head changes position

← Movement of actin filament

(d) The myosin head detaches from the actin filament as a result of an ATP molecule fixing to the myosin head

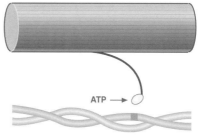

ATP →

(e) The ATP provides the energy to cause the myosin head to be 'cocked' again. The hydrolysis of the ATP gives rise to ADP + P

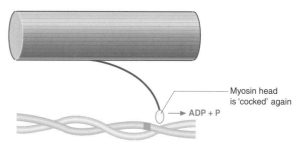

Myosin head is 'cocked' again

→ ADP + P

(f) The 'cocked' head of the myosin filament reattaches further along the actin filament and the cycle of events is repeated

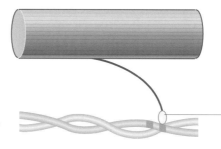

New point of attachment on actin filament

Fig 28.5 Sliding filament mechanism of muscle contraction

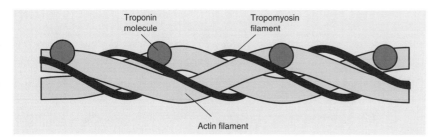

Troponin molecule

Tropomyosin filament

Actin filament

Fig 28.6 Relationship of tropomyosin and troponin to the actin filament

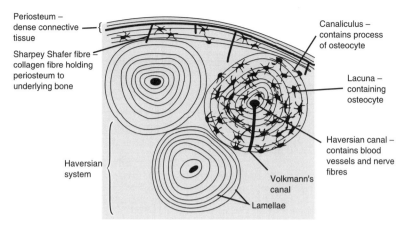

Periosteum – dense connective tissue

Sharpey Shafer fibre – collagen fibre holding periosteum to underlying bone

Haversian system

Canaliculus – contains process of osteocyte

Lacuna – containing osteocyte

Haversian canal – contains blood vessels and nerve fibres

Volkmann's canal

Lamellae

Fig 28.7 Compact bone (TS)

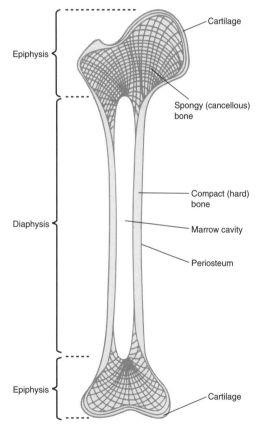

Epiphysis

Cartilage

Spongy (cancellous) bone

Compact (hard) bone

Marrow cavity

Periosteum

Diaphysis

Epiphysis

Cartilage

Fig 28.8 Vertical section through the femur

29 Control systems in plants

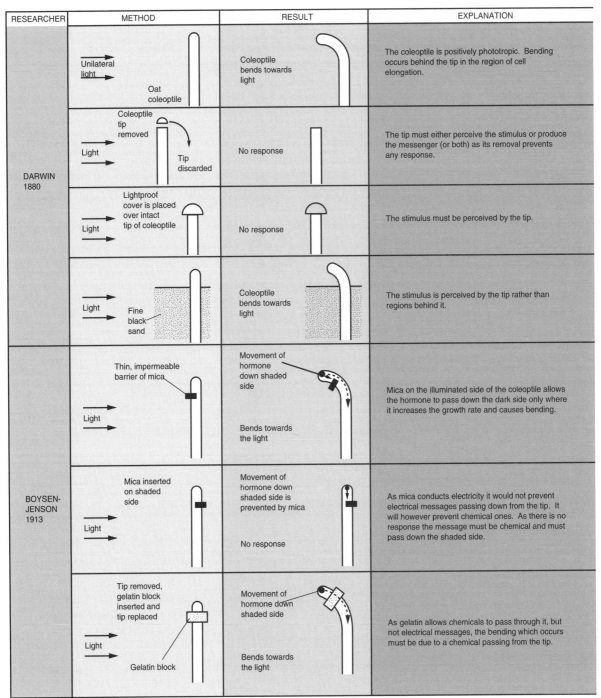

RESEARCHER	METHOD	RESULT	EXPLANATION
DARWIN 1880	Unilateral light — Oat coleoptile	Coleoptile bends towards light	The coleoptile is positively phototropic. Bending occurs behind the tip in the region of cell elongation.
	Coleoptile tip removed — Light — Tip discarded	No response	The tip must either perceive the stimulus or produce the messenger (or both) as its removal prevents any response.
	Lightproof cover is placed over intact tip of coleoptile — Light	No response	The stimulus must be perceived by the tip.
	Light — Fine black sand	Coleoptile bends towards light	The stimulus is perceived by the tip rather than regions behind it.
BOYSEN-JENSON 1913	Thin, impermeable barrier of mica — Light	Movement of hormone down shaded side — Bends towards the light	Mica on the illuminated side of the coleoptile allows the hormone to pass down the dark side only where it increases the growth rate and causes bending.
	Mica inserted on shaded side — Light	Movement of hormone down shaded side is prevented by mica — No response	As mica conducts electricity it would not prevent electrical messages passing down from the tip. It will however prevent chemical ones. As there is no response the message must be chemical and must pass down the shaded side.
	Tip removed, gelatin block inserted and tip replaced — Light — Gelatin block	Movement of hormone down shaded side — Bends towards the light	As gelatin allows chemicals to pass through it, but not electrical messages, the bending which occurs must be due to a chemical passing from the tip.

Fig 29.1 Diagrammatic summary of the historical events leading to the discovery of auxin and an understanding of its mechanism of action

29 Control systems in plants

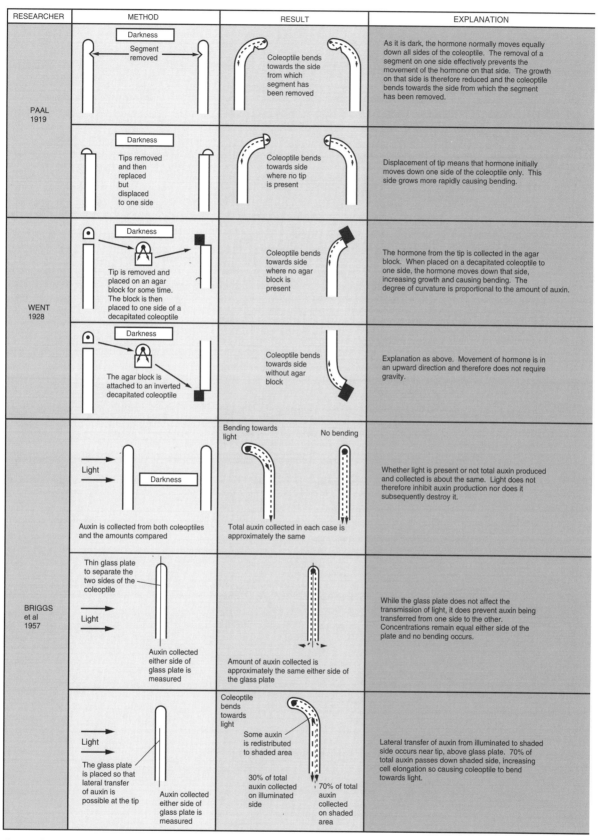

RESEARCHER	METHOD	RESULT	EXPLANATION
PAAL 1919	Darkness — Segment removed	Coleoptile bends towards the side from which segment has been removed	As it is dark, the hormone normally moves equally down all sides of the coleoptile. The removal of a segment on one side effectively prevents the movement of the hormone on that side. The growth on that side is therefore reduced and the coleoptile bends towards the side from which the segment has been removed.
	Darkness — Tips removed and then replaced but displaced to one side	Coleoptile bends towards side where no tip is present	Displacement of tip means that hormone initially moves down one side of the coleoptile only. This side grows more rapidly causing bending.
WENT 1928	Darkness — Tip is removed and placed on an agar block for some time. The block is then placed to one side of a decapitated coleoptile	Coleoptile bends towards side where no agar block is present	The hormone from the tip is collected in the agar block. When placed on a decapitated coleoptile to one side, the hormone moves down that side, increasing growth and causing bending. The degree of curvature is proportional to the amount of auxin.
	Darkness — The agar block is attached to an inverted decapitated coleoptile	Coleoptile bends towards side without agar block	Explanation as above. Movement of hormone is in an upward direction and therefore does not require gravity.
BRIGGS et al 1957	Light — Darkness — Auxin is collected from both coleoptiles and the amounts compared	Bending towards light / No bending — Total auxin collected in each case is approximately the same	Whether light is present or not total auxin produced and collected is about the same. Light does not therefore inhibit auxin production nor does it subsequently destroy it.
	Thin glass plate to separate the two sides of the coleoptile — Light — Auxin collected either side of glass plate is measured	Amount of auxin collected is approximately the same either side of the glass plate	While the glass plate does not affect the transmission of light, it does prevent auxin being transferred from one side to the other. Concentrations remain equal either side of the plate and no bending occurs.
	Light — The glass plate is placed so that lateral transfer of auxin is possible at the tip — Auxin collected either side of glass plate is measured	Coleoptile bends towards light — Some auxin is redistributed to shaded area — 30% of total auxin collected on illuminated side / 70% of total auxin collected on shaded area	Lateral transfer of auxin from illuminated to shaded side occurs near tip, above glass plate. 70% of total auxin passes down shaded side, increasing cell elongation so causing coleoptile to bend towards light.

Fig 29.1 (continued) Diagrammatic summary of the historical events leading to the discovery of auxin and an understanding of its mechanism of action

29 Control systems in plants

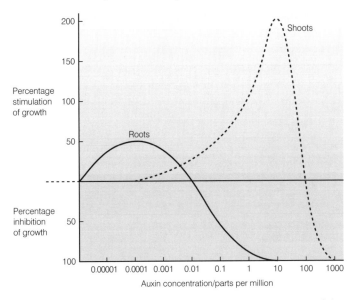

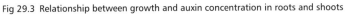

Fig 29.3 Relationship between growth and auxin concentration in roots and shoots

30 Biotechnology

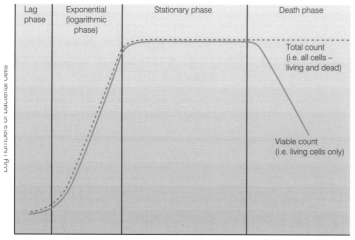

Fig 30.1 Bacterial growth curve

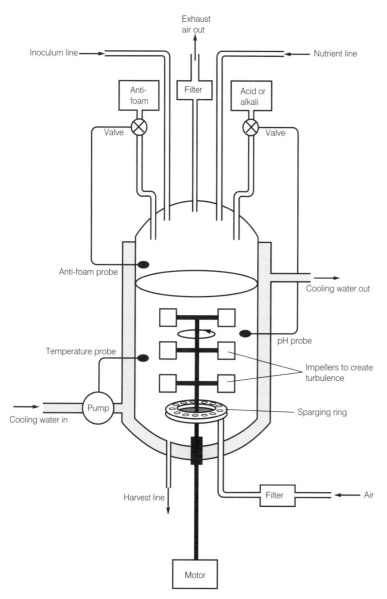

Fig 30.2 A stirred-tank fermenter

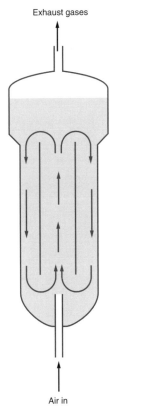

Fig 30.3(a) Air-lift fermenter

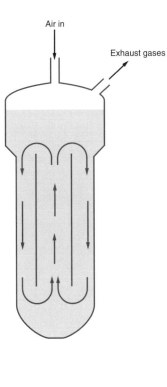

(b) Deep-shaft fermenter

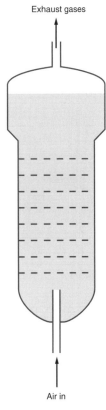

(c) Bubble-column fermenter

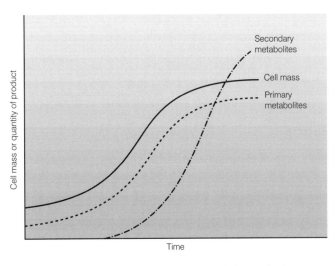

Fig 30.4 Comparison of primary and secondary metabolite production

31 Human health and disease

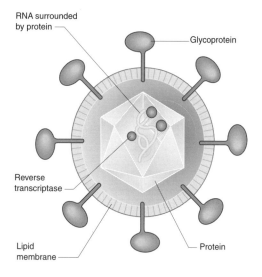

RNA surrounded by protein

Glycoprotein

Reverse transcriptase

Lipid membrane

Protein

Fig 31.1 The AIDS virus

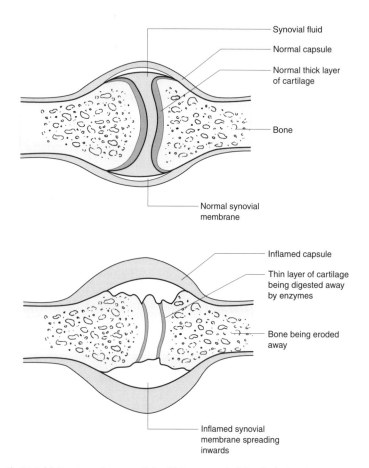

Synovial fluid

Normal capsule

Normal thick layer of cartilage

Bone

Normal synovial membrane

Inflamed capsule

Thin layer of cartilage being digested away by enzymes

Bone being eroded away

Inflamed synovial membrane spreading inwards

Fig 31.4 (a) Structure of a normal joint (b) Structure of a joint displaying rheumatoid arthritis